Membrane Separations in Chemical Engineering

A.E. Fouda, J.D. Hazlett,
Takeshi Matsuura and J. Johnson, editors
C.R. Bartels, B.S. Minhas and B. Ryan, co-editors

C.R. Bartels
E.R. Beaver
R.-D. Behling
J.G.A. Bitter
Michael S.K. Chen
Shihan Chen
K.J. Doshi
H.E. Ehresmann
K. Enomoto
Brian A. Farnand
A.E. Fouda
J.P. Haan
K. Haraya
J.D. Hazlett
M.W. Hellums
V.D. Helm
A. Higuchi
G.R. Husk
N. Itoh
Yuen-Koh Kao
W.J. Koros
O. Kutowy
E. Kyburz
Gregory S. Markiewicz
T. Matsuura
M.J. Mitariten
M.K. Murphy
T. Nakagawa
H. Nakano
Soo Hong Noh
K. Ohlrogge
M. Pasternak
D.R. Paul
K.-V. Peinemann
R. Rautenbach
J. Reale
A.W. Rice
H.C. Rijkens
Charles E. Rogers
V.M. Shah
Y. Shindo
T.A. Tweddle
Kamesh G. Venugopal
R.G. Werner
J.G. Wijmans
Zhiquan Yan
Tahua Yang
C. Yong

AIChE Symposium Series
Number 272 1989 Volume 85
Published by
American Institute of Chemical Engineers
345 East 47 Street New York, New York 10017

American Institute of Chemical Engineers
345 East 47 Street, New York, N.Y. 10017

AIChE shall not be responsible for statements or opinions advanced in papers or printed in its publications.

Library of Congress Cataloging-in-Publication Data

Membrane separations in chemical industries.

(AIChE symposium series ; no. 272, v. 85)
Papers presented at the AIChE 1989 Houston National Meeting.

1. Membrane separation—Congresses. I. Matsuura, Takeshi, 1936- . II. National Meeting of AIChE (1989 ; Houston, Tex.). III. Series: AIChE symposium series ; no. 272.
TP156.S45M46 1989 660′.2842 89-17963
ISBN 0-8169-0482-0

FOREWORD

This symposium volume includes papers presented in three sessions on Membrane Gas Separation (Membrane Gas Treatment I, II and III) and one session on Membrane Separations for Fuel and Hydrocarbon Applications at the 1989 AIChE Spring National Meeting, Houston, April 2–6. These technical sessions were jointly sponsored by the CSChE and AIChE, with American and Canadian researchers responsible for the organization and much of the content of these sessions.

The first chapter of the volume deals with the membrane development and the membrane post-treatment for various gas separation processes. Membrane gas separation process designs and also novel process designs in which membrane gas separations are important components is the focus of the second chapter The third chapter is on membrane applications in the petroleum and petrochemical industries, particularly separations of nonaqueous solutions. Some papers are rich in fundamentals and others are more concerned with applications and, in the volume as a whole, both aspects should be well balanced.

All editors and co-editors of this volume served as session chairpersons for the sessions described above. They are: A.E. Fouda, J.D. Hazlett, T. Matsuura of National Research Council Canada, J. Johnson of Chevron Corporation, C.R. Bartels of Texaco, Inc., B.S. Minhas of W.R. Grace and Co., and B. Ryan of United Engineers and Constructors, Inc.

The editors would like to express their sincere appreciation to the contributors of papers for their effort in manuscript preparation. They also wish to thank Maura Mullen and Maryanne Spencer of the American Institute of Chemical Engineers for their invaluable editorial assistance.

A.E. Fouda, J.D. Hazlett, Takeshi Matsuura,
and J. Johnson, *editors*

CONTENTS

THE HIGHEST GAS PERMEABLE MEMBRANE OF POLY[1-(TRIMETHYLSILYL)-1-PROPYNE] MODIFIED BY FILLING POLYORGANOSILOXANE

T. Nakagawa, H. Nakano, K. Enomoto and A. Higuchi ■ Department of Industrial Chemistry, Meiji University
Higashi-mita, Tama-ku, Kawasaki 214 Japan

Poly[1-(trimethylsilyl)-1-propyne] (PMSP) membrane has the highest gas permeability. This property, however, decreased with time or thermal hysteresis. This paper shows effects of poly(dimethylsiloxane) or poly(trifluoropropylmethylsiloxane) filled in the PMSP membrane on the diffusivity, solubility and permeability, especially the stability of gas permeability.

INTRODUCTION

In recent years, considerable work was done on gas separation by membranes (1). The most important properties of membranes used for gas separation are gas permeability, separation factor and durability of the membrane. The utility of polymeric membranes for separating a gas mixture was known even in 19th century (2). However, due to the low gas permeability, membrane separation technology was not used as an industrial scale in this field. Until a new polymer, poly[1-(trimethylsilyl)-1-propyne], PMSP, was synthesized by Masuda and Higashimura (3,4,5), poly(dimethyl siloxane), PDMS, was the highest gas permeable membrane. The permeability coefficient of PDMS to oxygen, for example, has been reported to be 6.0 x 10^{-8} cm^3(STP) cm cm^{-2} s^{-1} $cmHg^{-1}$ at 30 C (6), which differs depending on the degree of crosslinking. However, it was found that the permeability coefficient of PMSP to oxygen is 10~50 times higher than that of PDMS (7).

Recently Odani and his coworkers reported a noticeable physical aging effect on permeability coefficients of this polymer. For example, permeability coefficients for iso-butane and oxygen at 30°C decreased by about two magnitudes and one magnitude, respectively, over the period of 100 days under vacuum (8). Authors found that permeability coefficients of this polymer to hydrogen, oxygen and nitrogen also decreased remarkably by thermal hysteresis or adsorption of less volatile vapors. These phenomena seem to be a serious defect for a practical use. In order to stabilize the gas permeability of PMSP, one of the authors has tried some methods of modification, and reported a modification by adsorption of very small amount of plasticizers such as dioctylphtharate, DOP (9). The objective of this investigation was to study the effect of filling of liquid PDMS or oligomer of fluorine containing polyorganosiloxane on gas permeability coefficients and especially the stability of the permeability.

EXPERIMENTAL

Preparation of membranes

Homogeneous PMSP membranes were prepared in our laboratory by dissolving PMSP, which was synthesized using $TaCl_5$ as a catalyst according to Masuda's method (3), in a toluene, and then casting a thin film of solution on a glass plate and dried under vacuum. Membranes were immersed in a methanol to keep the membranes. The chemical structure of PMSP is as follows:

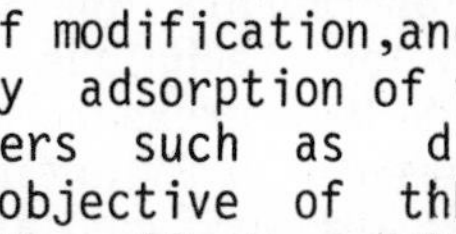

Modification of membranes

Two kinds of filling materials were used: a liquid PDMS, KF-96 1000 CS, whose average molecular weight was 2.6 x 10^4 was supplied from shinetsu Polymer Co. and poly 1,3,5-tris (3,3,3-trifluoropropyl)-1,3,5-trimethyl cyclotrisiloxane, PTFMS, whose molecular weight was 1.6×10^3, synthesized by ourselves from a corresponding monomer. PMSP membranes were immersed in a solution of methlethyl ketone dissolved the liquid PDMS or PTFMS by appropriate concentrations at room temperature for about two days, followed by being taken out from the solution, and then wiped and dried under vacuum. The ammount of the liquid PDMS or PTFMS filled in the membrane was decided by a weight difference of the membrane before and after immersion. The chemical structure of PTFMS is as follows:

$$-\!\!\left(\begin{array}{c} CH_3 \\ | \\ Si\text{-}O \\ | \\ CH_2CH_2CF_3 \end{array}\right)\!\!-$$

Measurements of gas solubility and gas permeability

The whole apparatus was kept at 35°C. Elongations of a quartz spring by absorptions of gases was monitored with a detector and recorded. The general theory of gas transport in polymers and detailed discussions of the methods of measurement and calculation of the permeability, diffusivity and solubility coefficients have been published elsewhere (10). The experimental method used in this study was an adaptation of a high vacuum gas transmission technique with MKS Baratron Model 310HS-100S pressure transducer (9). In our experiment, a Silastic silicone rubber with no additives was used as a rubber gasket.

X-ray and other characterization

The method used for X-ray diffraction characterization was carried out as usual and a viscoelastometer, Rheovibron DDV-II-C, manufactured by Toyo Boldwin Co. was used for viscoelastic properties.

RESULTS AND DISCUSSION

Effect of liquid PDMS and PTFMS filled in PMSP membranes on their molphology

X-ray analysis of PMSP membranes filled with PDMS and PTFMS are shown in Figs.1 and 2, respectively. Both fillers decrease height of sharp peak at 10° of 2θ. Comparing X-ray patterns, the effect of PTFMS seems to be stronger than that of PDMS, namely less amount of PTFMS gives decrease of the hight. The effect of 5-6 wt% PDMS and PTFMS filled in the membranes on the viscoelastic properties of E' and tan is shown in Fig.3. Although E' values are almost the same for unmodified and modified membranes, the effect of PTFMS on tanδ is quite different with PDMS, which shows a new peak about 90°C. This suggests a new morphology in the modified membrane.

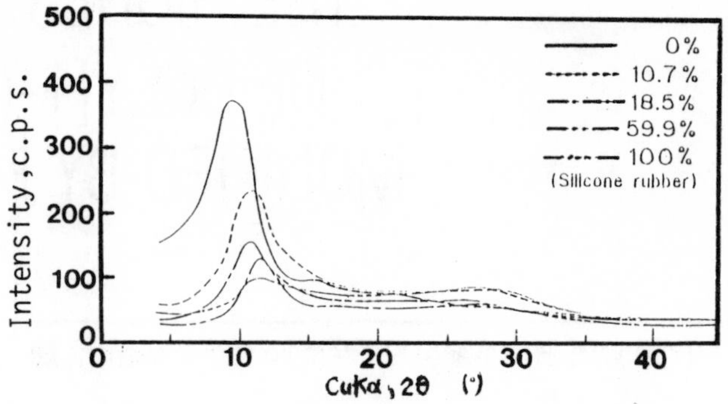

Fig.1. X-ray diffraction patterns of PMSP membranes filled with PDMS

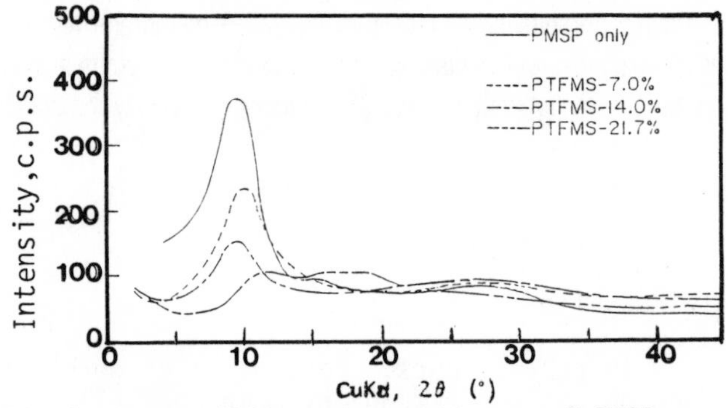

Fig.2. X-ray diffraction patterns of PMSP membranes filled with PTFMS

Effect of liquid PDMS and PTFMS filled in PMSP membranes on S, D and P

One of the main reasons for the unstability in gas permeation property of PMSP membrane is considered to be an unrelaxed large excess free volume (11), which is very easy to reduce. Our idea for stabilization of gas permeabilities of PMSP is to control too much larger excess free volume by filling of small amounts of liquid state polymer or oligomer and thus to obtain the modified PMSP membranes which have the permeability at limitted reduced permeability coefficients.

In order to clearify the effect of the filling materials on S, D and P. Arrhenius plots of P, S and D of CO_2, H_2, CH_4, O_2 and N_2 in the unmodified PMSP are shown in Fig.4. Fig. 4 shows that the PMSP membrane is nonporous because of positive E_D values which means an activated diffusion process. PMSP has very large S, which decreases remarkably with increasing temperature. Quantitative description of solubility of a gas in a glassy polymer was achieved by Michel, Vieth and Barrie (12), and has become known as the Dual Sorption Theory. The equilibrium part of the theory is simply expressed by the following equation for the isotherm:

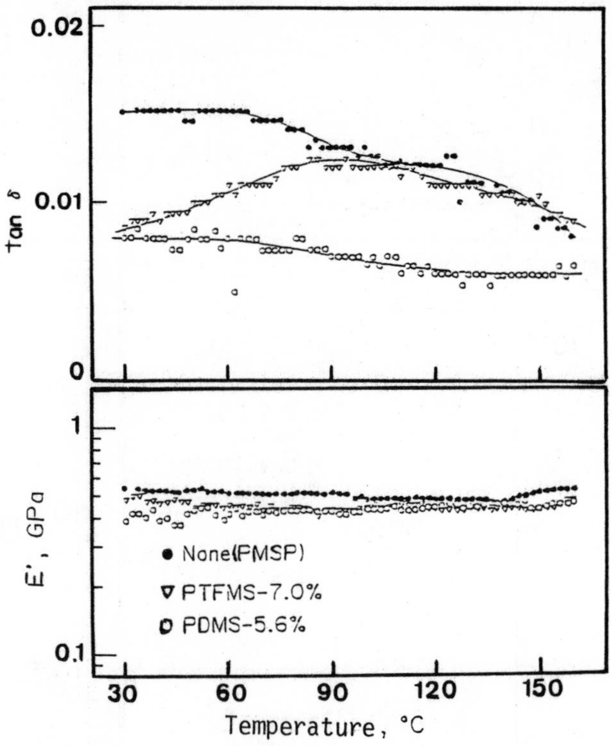

Fig.3. Dinamic viscoelasticity of PMSP membranes containing PDMS and PTFMS

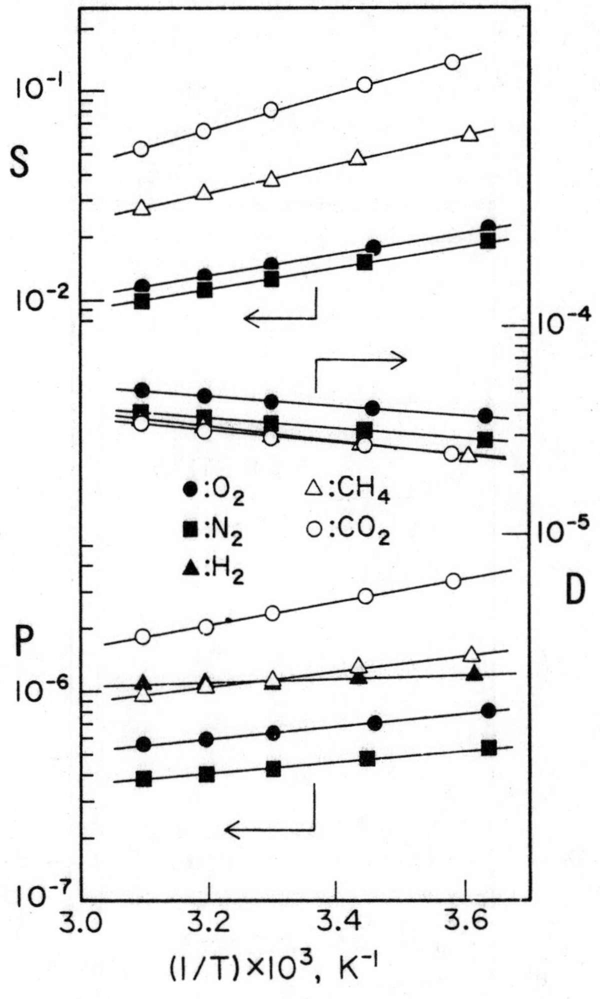

Fig.4. Temperature dependence of S, D and P of gases to PMSP membrane

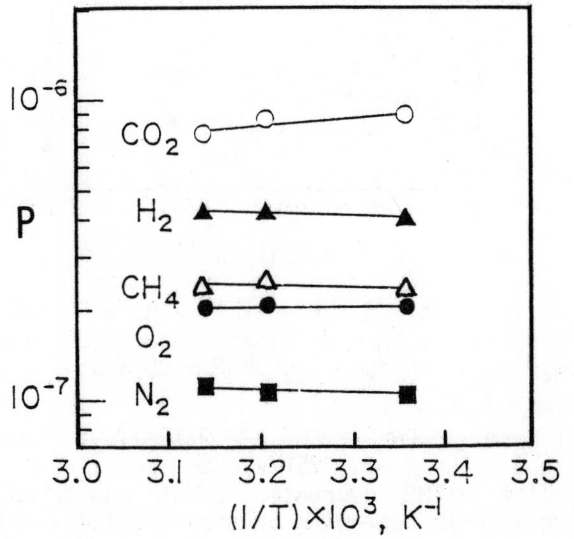

Fig.5. Temperature dependence of permeabilities of gases to PMSP/PTFMS (87.9/12.1) membrane

$$C = C_D + C_H = k_D p + \frac{C'_H bp}{1 + bp} \qquad (1)$$

$$P = k_D D_D + \frac{D_H \cdot C'_H \cdot bp}{1 + bp} \qquad (2)$$

This theory hypothesizes that glassy polymer has two slightly different molecular environments: one that is dence where gas molecules can dissolve by obeying Henry's law, and one that is composed of uniformly distributed molecular-scale cavities and absorb on their walls according to the Langmuir isotherm. One of the authors has already reported that glassy PMSP has a very large excess free volume, particularly in the Langmuir domains in the polymer.The latest Dual-mode parameters of CO_2 for the PMSP membrane at 35°C are as follows: k_D= 0.886, C'_H= 139.7, b= 0.0469, D_D= 15.9 x 10^{-5}, D_H= 2.62 x 10^{-5}.

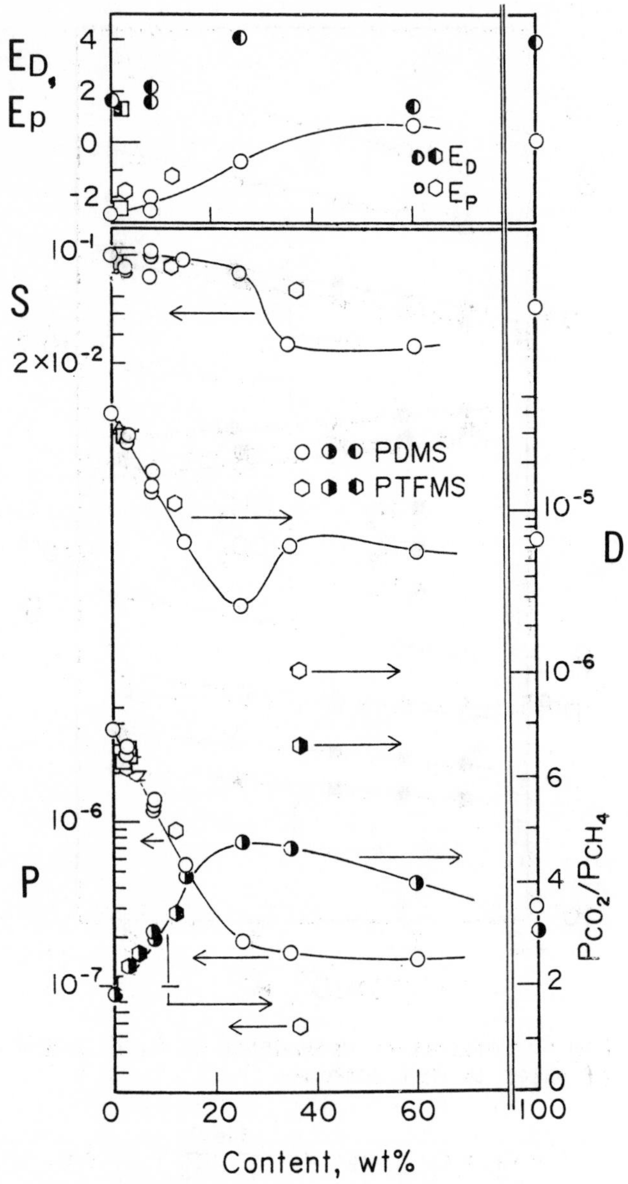

Fig.6. Effect of contents of PDMS or PTFMS on P, D, S, E_P, E_D of CO_2 and separation factors, P_{CO_2}/P_{CH_4}, to modified PMSP membranes

Considering the results of the X-ray analysis and viscoelastic characterizations, the liquid PDMS might be mainly filled in the Langmuir's domain and partly in the dence region, whereas PTFMS might be present in the dence region and mixed well with polymer segments, because an affinity of the fluorine containing polyorganosiloxane with PMSP seems to be better than that of PDMS. Small amounts of the fillers occupy the Langmuir's domain and increase the gas solubility.

Temperature dependence of permeabilities

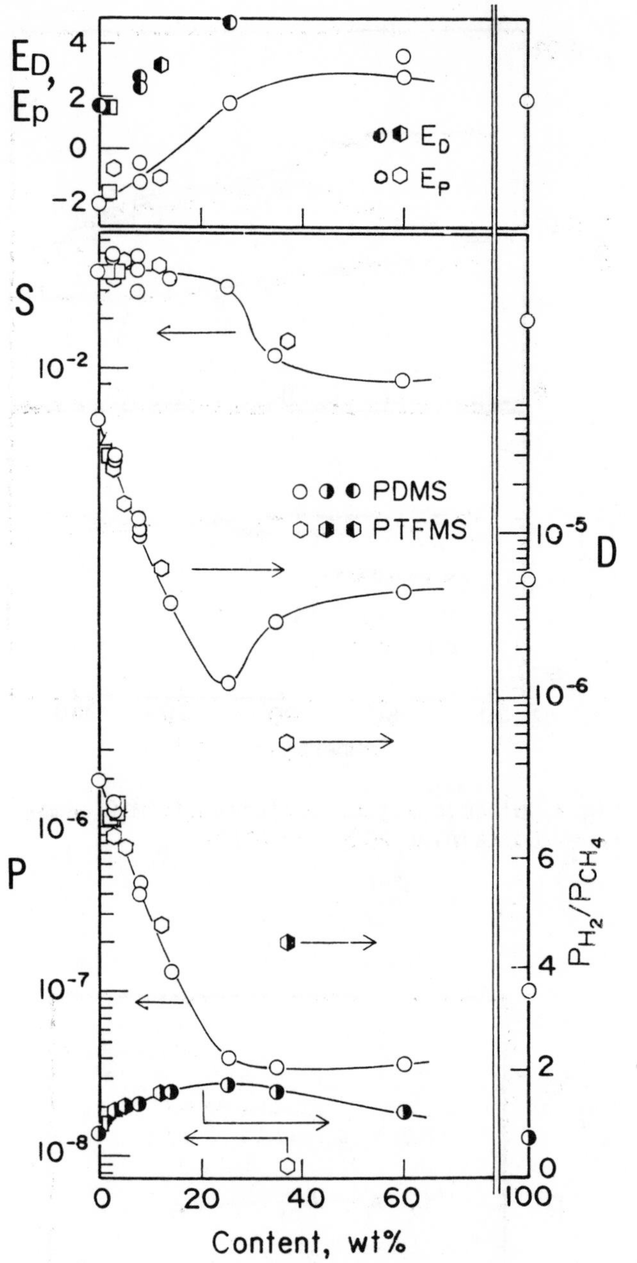

Fig.7. Effect of contents of PDMS or PTFMS on P, D, S, E_P, E_D of CH_4 and separation factors, P_{H_2}/P_{CH_4}, to modified PMSP membranes

of gases to PMSP/PTFMS (87.9/12.1) is shown in Fig.5. The P of PMSP/PTFMS containing 12 wt% of PTFMS decrease about one-third ~ one-seventh depending on gases. Also, inclines of these Arrhenius plots became flat, which means an increase of E_P. The same figure as Fig.5 was also obtained for PMSP/PDMS containing 12 wt% of PDMS. Effect of contents of PDMS and PTFMS in PMSP membranes on the P, D, S, E_P, E_D of methane, and separation factor

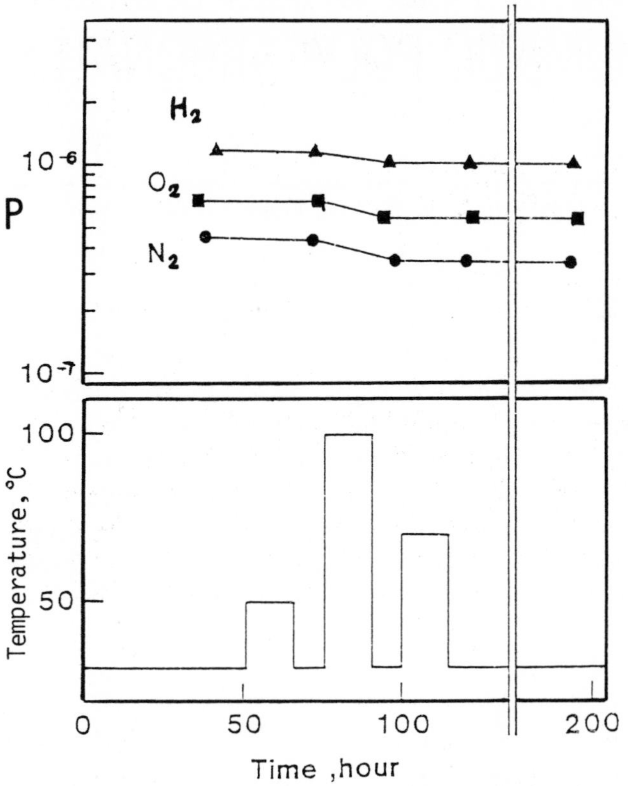

Fig.8. Effect of thermal hysteresis on P of PMSP/PDMS (99.3/0.7) and PMSP membranes to H_2, O_2 and N_2 at 30°C

P_{CO2}/P_{CH4}, are shown in Fig.6. Those of CH_4 and the separation factors, P_{H2}/P_{CH4}, are also shown in Fig. 7. P and D decreased with increasing the content in the region between 0 and 25 wt%, whereas E_P and E_D increased. The effect of higher content of PDMS or PTFMS, however, seems to diminish. Perhaps, P, D of the modified PMSP membranes might reach to intrinsic values of poly(dimethyl siloxane) or poly(trifluoropropylmethyl siloxane).

Effect of hysteresis on P of H_2, O_2 and N_2 at 30°C to PMSP and PMSP/PDMS or PMSP/ PTFMS membranes was examined. One of the results is shown in Fig. 8. The thermal stability of the modified PMSP membrane filled with 0.7 wt% of PDMS was found to be excellent. It is considered that the small amount of PDMS covers "inner surface" of Langmuir's domains, and protect "collapse" of the unrelaxed excess free volume and stabilize the permeability.

NOTATION

- b hole affinity constsnt, cc $cmHg^{-1}$
- C concentration, cc gas(STP)/cc polymer
- C_D concentration by normal dissolution, cc gas (STP)/cc polymer
- C_H concentration by hole-filling, cc gas(STP)/ cc polymer
- C'_H hole saturation coefficient, cc gas(STP)/cc polymer
- D apparent diffusion coefficient, cm^2/sec
- D_D diffusion coefficient in Henry's domain, cm^2/sec
- D_H diffusion coefficient in Langmuir's domain, cm^2/sec
- E_D apparent activation energy for diffusion, kcal/mole
- E_P apparent activation energy for permeation, kcal/mole
- k_D Henry's law solubility coefficient, cc gas (STP)/cc polymer cmHg
- P permeability coefficient, cc(STP)cm/cm^2 sec cmHg
- p supplied pressure of gas, cmHg
- S apparent solubility coefficient, cc gas(STP) /cc polymer cmHg

LITERATURE CITED

1. Lonsdale, H.K., J. Memb. Sci.10, 81(1982).
2. Mitchell, J.V.,J. Roy.Inst. Great Britain, 2, 101, 307 (1831).
3. Masuda, T., E. Isobe, T. Higashimura, and K. Takada, J. Am. Chem. Soc.,105, 7473 (1983).
4. Masuda, T.,T. Higashimura, Acc.Chem.,Rev., 17, 51 (1984).
5. Masuda, T., E. Isobe, and T. Higashimura, Macromolecule, 18, 841 (1985).
6. Ward III, W. J., W. R. Browall, and R. M. Salemme, J. Memb. Sci., 1, 99 (1976).
7. Takada, K., H. Matsuya, T. Masuda, and T. Higashimura, J. Appl. Polym. Sci., 30, 1605 (1985).
8. Shimomura, H., K. Nakanishi, H. Odani, M. Kurata, T. Masuda, and T. Higashimura, Kobunshi Ronbunshu, 43, 747 (1986).
9. Nakagawa, T.,T. Saito, S. Asakawa, and Y. Saito, Gas Sepr. & Purification, 2,3(1988).
10. Nakagawa, T., H. B. Hopfenberg, and V. T. Stannett, J. Appl. Polym. Sci., 15, 231 (1971).
11. Ichiraku,Y.,S.A. Stern, and T. Nakagawa, J. Memb. Sci., 34, 5 (1987).
12. Michaels, A. S., W. R. Vieth, and J. A. Barrie, J. Appl. Phys., 34, 1 (1963).

GAS PERMEABILITY AND CHAIN PACKING IN AROMATIC POLYCARBONATES

M.W. Hellums, W.J. Koros and D.R. Paul ■ Department of Chemical Engineering, University of Texas, Austin, TX 78712-1062
G.R. Husk ■ U.S. Army Research Office, Research Triangle Park, NC 27709-2211

The gas transport properties in a series of four polycarbonates is characterized for CO_2, CH_4, O_2, and N_2. The permeability of gases in the highest flux material, Tetramethylhexafluorobisphenol-A polycarbonate is roughly fifteen times higher than in conventional PC. Mixed gas permeation results indicate plasticization tends to lower permselectivity for the CO_2/CH_4 pair below the values from pure gas data. Wide angle x-ray diffraction measurements of the average segmental spacings in the series of polycarbonates reflect hindrance to chain packing caused by variations in structure. Relationships between dynamic-mechanical property transitions and the permselectivity are proposed. Hindrance to intrachain rotational mobility is apparent from the sub-Tg transition temperatures measurements in the substituted materials. Support is given to the concept that simultaneous inhibition of chain packing and inhibition of intrachain rotational mobility can lead to improved membrane performance.

INTRODUCTION

Studies of the relationships between polymer structures and transport properties have shown promise in demonstrating significant improvements in the intrinsic separation performance of membrane materials. Typically a trade off exists between gas permeability and permselectivity in polymeric materials as shown in Figure 1. However, recent studies [1, 2, 3, 4] have indicated that systematic variations of the chemical structure can be used to achieve simultaneous increases in permeability and permselectivity. The present study examines the effects of systematic variations in the chemical structure on the gas permeability and chain packing in substituted polycarbonates. In addition, relationships between transport properties and dynamic mechanical properties are examined.

BACKGROUND AND THEORY

A key index of the productivity of a membrane material is the permeability, P. The permeability of a gaseous penetrant, can be written as a simple product of an average diffusivity, D, and an effective solubility, S, of the penetrant in the polymer matrix [5, 6].

$$P = D\,S \qquad (1)$$

For conditions of negligible downstream pressure, the solubility coefficient, S, is equivalent to the secant slope of the gas sorption isotherm evaluated at the upstream conditions. The average diffusivity, D, provides a measure of the effective mobility of the penetrant in the polymer matrix between the conditions at the upstream and the downstream side of the film.

The second key index of the performance of a membrane material is the separation factor, or permselectivity, $\alpha_{A/B}$. When the downstream pressure is negligible, $\alpha_{A/B}$ is rigorously equal to the ideal separation factor, $\alpha^*_{A/B}$. In the absence of plasticizing responses due to strong gas-polymer interactions, $\alpha^*_{A/B}$ can be approximated using the ratio of permeabilities for pure gases A & B [7].:

$$\alpha^*_{A/B} = \frac{P_A}{P_B} \qquad (3)$$

If Eq (1) is substituted into Eq (2), the ideal separation factor can

$$\alpha^*_{A/B} = \left[\frac{D_A}{D_B}\right]\left[\frac{S_A}{S_B}\right] \qquad (4)$$

where D_A/D_B is the diffusivity selectivity, and S_A/S_B is the solubility selectivity. The solubility selectivity is determined by the differences in condensibility of the two penetrants and by their interactions with the membrane material. The diffusivity selectivity is based on the inherent ability of polymer matrices to function as size and shape selective media through segmental mobility and intersegmental packing factors.

As shown in Figure 1 some notable exceptions to the typical trade-off in permeability and permselectivity were demonstrated in the polyimide family [3] and in the present study of polycarbonates. Certain structural variations were found to produce very substantial increases in permeability and permselectivity. The findings of the earlier studies, supported by the present results on the polycarbonates, can be generalized into principles for membrane material selection:

(1) Substitutions which inhibit chain packing and thereby open-up the polymer matrix lead to increases in permeability and decreases in permselectivity. Substitutions which restrict rotational mobility of the chains about the flexible linkages on the polymer backbone lead to increases in permselectivity and decreases in permeability.

(2) A beneficial compromise can be found by ***simultaneously*** inhibiting chain packing and rotational mobility about flexible linkages which can lead to an increase in both permeability and permselectivity.

(3) Substitutions that decrease the concentration of mobile linkages in the polymer backbone tend to lead to increased permselectivity without a large reduction in permeability.

The data from the current study of the polycarbonate family is in substantial agreement with these principles as discussed below.

EXPERIMENTAL

The polycarbonates studied here are shown in Figure 2. The reference material for the study is conventional bisphenol-A polycarbonate, PC, obtained from the General Electric Company. The TMPC and HFPC were experimental polymers supplied by the Dow Chemical Company. The TMHF-PC was synthesized by the authors using an interfacial polymerization method [8, 9]. The films used in the study were all cast from methylene chloride solutions. The permeabilities of nitrogen, oxygen, methane, and carbon dioxide were measured using the standard permeation apparatus employed in our labs. The gas sorption measurements were carried out with a pressure decay sorption cell [10]. All the permeation and sorption measurements were made at 35 ^{o}C.

Wide angle x-ray diffraction (WAXD) measurements were performed to characterize the polymer intersegmental spacings. A Philips x-ray diffractometer was used with Cu K-α radiation, wavelength = 1.54 Å. Each of the materials produced broad peaks characteristic of amorphous materials. The angle at the center of the peak on each x-ray diffraction pattern is representative of the average intersegmental spacing of the polymer backbone centerline. The intersegmental spacing, or d-spacing, can be calculated with Bragg's equation [11], $n\lambda=2d\sin\theta$. The d-spacings are shown in Table 1.

The glass transition temperatures (Tg) measured by Differential Scanning Calorimetry at a heating rate of 20 ^{o}C per minute and the sub-Tg transition temperatures (T_γ) were used as measures of the chain "stiffness" or rotational mobility. The T_γ data tabulated below for PC, HFPC, and TMPC are from the literature [12]. The T_γ of TMHF-PC was measured using a Rheometrics dynamic mechanical tester.

RESULTS AND DISCUSSION

The data show that the higher permeabilities and permselectivities of the rigid substituted polycarbonates occur due to both solubility and diffusivity factors. For example, dramatically higher gas diffusivities are apparent when comparing the very open TMHF-PC to conventional PC. The higher sorption levels in the substituted polycarbonates relative to PC primarily reflect the increased free space in the materials which makes it energetically easier to accommodate a sorbed penetrant.

Table 1. Polycarbonate Characterizations

	PC	TMPC	HFPC	TMHF-PC
Tg, (^{o}C)	150	193	176	208
T_γ, (^{o}C)	-100	50	-100	50
d-space, (Å)	5.2	6.0	5.8	6.3
density, (g/cc)	1.200	1.083	1.478	1.286
Fractional Free Volume (V_f/V_p)	0.164	0.180	0.196	0.216
Conc. Carbonyl Groups (µmol/cc)	4720	3460	4080	3070

The fluorine-containing polycarbonates exhibit significantly higher permeabilities relative to the corresponding non-fluorinated polycarbonates (HFPC vs PC and TMHF-PC vs TMPC) and either higher or only slightly lower permselectivities depending on the gas pair (see Tables 2-3).

For this series of materials the x-ray d-spacings and the fractional free volumes provide a useful measure of the extent of "openness" of the polymers. As shown in Figures 3 and 4 the gas permeabilities are generally higher in materials with higher x-ray d-spacings and higher estimated fractional free volume. The density does not correlate well with permeability in this series of materials.

The mobility of the polymer chains also plays a role in the permeability and permselectivity. The d-spacing difference between HFPC and TMPC is only 0.2 Å and the large difference in the T_γ of the two materials indicates significant differences in intrasegmental motions. As discussed below this difference in intrasegmental motions between HFPC and TMPC may explain why TMPC has a lower permeability than HFPC despite its higher d-spacing.

Studies of the time scale and temperature dependence of mechanical properties [11, 13] and pulsed and wideline NMR studies [14, 15] of polycarbonates have indicated that the gamma transition involves small scale rocking motions of single repeat units about the carbonate linkage. At temperatures below T_γ, these torsional motions are greatly suppressed. Both TMPC and TMHF-PC have a T_γ of 50 ^{o}C vs. -100 ^{o}C for PC and HFPC. This large difference in T_γ between the materials with and without the tetramethyl-substitution suggests that the segmental rotations about the carbonate linkage are hindered by the tetramethyl-substitution.

The suppression of chain mobility is believed to narrow the size distribution of gaps that open and close in the medium to allow a diffusive jump by a gas molecule. This should then cause a decrease in the gas diffusion coefficient (lowering permeability). Larger molecules should be more strongly affected than smaller ones - leading to an increase in diffusivity selectivity. This is presumably the reason that the diffusivity selectivities for TMPC are higher than for PC even though the matrix of TMPC is more open and more permeable. Similarly, the diffusivity selectivities of TMHF-PC are as high as those of HFPC due to the mobility restriction offered by the tetramethyl substitution, even though the TMHF-PC is considerably more permeable than HFPC.

The solubility selectivities for CO_2/CH_4 have been shown to increase in polymers with increasing concentration of polar groups [16]. The solubility and diffusivity selectivities are plotted versus the carbonyl group concentration in Figure 5. It is interesting to note that the diffusivity selectivities for CO_2 and CH_4 are an even stronger function of the carbonyl group concentration than the solubility selectivities. As noted above, it is believed that reducing the concentration of flexible linkages should improve diffusivity selectivity. Naturally any substitution which dilutes the flexible carbonate linkage in a polycarbonate dilutes the density of carbonyl groups by the same amount. Therefore, increases in diffusivity selectivity which may occur from the dilution of carbonate group density will tend to be accompanied by decreases in solubility selectivity (at least for CO_2/CH_4).

The solubility selectivities for fluorine-containing polycarbonates fall above the average line for the series suggesting the existence of some interaction that may increase the CO_2 solubility. We also observed that CO_2 permeabilities of both of the fluorinated materials increased sharply at 15 to 25 atm of CO_2, indicating the onset of plasticization. Pure and mixed gas permeabilities for CO_2 and CH_4 in TMHF-PC are shown in Figures 6 and 7. The plasticization phenomena shown here may be due to dipolar interactions with CO_2 and the polar $-CF_3$ groups. This apparent interaction seems to improve the permselectivity when examining pure gas data. In fact, the mixed gas permeation experiments demonstrated that the CO_2/CH_4 permselectivity decreases with increasing CO_2 partial pressure (Figure 8). This behavior may limit the range of CO_2 partial pressure where the fluorinated materials could be used in membrane applications. Consequently this is an important example in which mixed gas permeation data are needed to reveal the true performance of a membrane material.

Table 2 - Permeabilities, Solubilities, and Diffusivities of CO_2 and CH_4 at 35 °C and 10 atm

	PC	TMPC	HFPC	TMHF-PC
P_{CO_2} (Barrers)	6.8	19	24	110
$\frac{P_{CO_2}}{P_{CH_4}}$	19	21	23	24
S_{CO_2} (ccSTP/cc atm)	1.6	2.6	2.7	4.1
$\frac{S_{CO_2}}{S_{CH_4}}$	4.0	3.2	3.9	3.4
D_{CO_2} ($10^{-8}cm^2/s$)	3.2	5.4	6.7	21
$\frac{D_{CO2}}{D_{CH4}}$	4.7	6.6	5.9	7.0

Table 3 - Permeabilities, Solubilities, and Diffusivities of O_2 and N_2 at 35 °C and 2 atm

	PC	TMPC	HFPC	TMHF-PC
P_{O_2} (Barrers)	1.6	5.6	6.9	32
$\frac{P_{O_2}}{P_{N_2}}$	4.8	5.1	4.1	4.1
S_{O_2} (ccSTP/cc atm)	0.21	0.46	0.47	0.78
$\frac{S_{O_2}}{S_{N_2}}$	1.5	1.2	1.3	1.3
D_{O_2} ($10^{-8}cm^2/s$)	7.9	9.2	11	31
$\frac{D_{O_2}}{D_{N_2}}$	3.2	4.1	3.2	3.2

CONCLUSIONS

TMHF-PC has the packing disruptions and mobility restrictions offered by both the tetramethyl- and hexafluoro-substitutions. As discussed above, each of these substitutions produce desirable changes in membrane properties. Consequently, it

The TMHF-PC has the largest x-ray d-spacings and the largest fractional free volume in the series. Consequently, TMHF-PC exhibits a dramatic increase in permeability relative to PC with a permeability to O_2 of 32 Barrers and a permeability to CO_2 of 110 Barrers. The high productivity coupled with a permselectivity to O_2/N_2 of 4.1 is a very good combination compared to commercially used materials.

This study of a homologous series of polycarbonates has added additional support to the notion that simultaneous inhibition of chain packing and restriction of intrachain torsional mobility can lead to improved gas separation membrane performance. Substantial increases in permeability have been achieved while maintaining permselectivity. A relationship between sub-Tg mechanical property transitions ($T\gamma$) and permselectivity is proposed. We have observed that increasing the rigidity of the polymer chain as well as restricting rotational mobility serves to improve material selectivity even when these substitutions substantially increase the permeability by increasing the average distance between polymer chains.

Of the four polycarbonates studied, the material which is both the most rigid as evidenced by its T_g, and the most open as evidenced by its x-ray d-spacing has an excellent combination of permeability and permselectivity. Both substitutions evaluated here produce substantial increases in permeability by hindering interchain packing. They also serve to increase permselectivity primarily by reducing intrachain rotational mobility which improves the size and shape discriminating ability of the matrix.

Although the substitutions primarily affect the permselectivity through the diffusivity component, solubility effects are also significant. Higher densities of carbonyl groups as well as the presence of trifluoromethyl groups tend to slightly increase the solubility selectivity of the CO_2/CH_4 and O_2/N_2 gas pairs. However, the high solubility of CO_2 produces a plasticization of the fluorinated materials causing decrease in mixed gas permselectivity at 15 to 25 atmospheres.

Measurements of x-ray d-spacing and estimates of fractional free volume from group contributions are good indications of the relative permeabilities of the materials. These techniques are useful in screening materials as candidates for membrane studies.

ACKNOWLEDGEMENT

The support of the Department of Energy, Basic Energy Sciences Program under research grant DE-F605-86ER13507 is acknowledged. The assistance of Dr. K.C. O'Brien of Dow Chemical for measuring the Tγ of TMHF-PC is also acknowledged.

LITERATURE CITED

1. H.H. Hoehn, Chapter 4 in *Material Science of Synthetic Membranes*, ACS Symposium Series, 269 (1985), D.R. Lloyd, Ed.

2. N. Muruganandam, W.J Koros, and D.R. Paul, *J. Polym. Sci: Part B: Polym. Phys.* 25, 1999 (1987)

3. T.H. Kim, W.J. Koros, G.R. Husk, K.C. O'Brien, *J .Memb. Sci.*, **37**, 45 (1988)

4. M.W. Hellums, W.J. Koros, G.R. Husk, D.R. Paul, "Fluorinated Polycarbonates for Gas Separation Applications", *J. Memb. Sci.*, in press (1989)

5. R.T. Chern, W.J. Koros, H.B. Hopfenberg, and V.T. Stannett, in *Material Science of Synthetic Membranes*, chapt. 2, ACS Symposium Series, 269 (1985), D.R. Lloyd, Ed.

6. W.J. Koros, G.K. Fleming, S. M. Jordan, T.H. Kim, H.H. Hoehn, *Prog. Poly.Sci.* , **13**, 339 (1988)

7. K.C. O'Brien, W.J. Koros, T.A. Barbari, and E.S. Sanders, *J. Memb. Sci.*, **29**, 229 (1986)

8. US Patent 4358624, Mark,V. and Hedges, C.V., assigned to General Electric Co., Mt. Vernon, New York (1982)

9. US Patent 3879348, V. Serini, H. Schnell, and Vernaleken , assigned to Bayer Aktiengsellschaft, Leverkusen, Germany (1975)

10. W.J. Koros, and D.R. Paul, *J. Polym. Sci: Polym. Phys. Ed.*, **14**, 1903 (1976)

11. L.H. Schwarts and J.B. Cohen, Chap. 3 in *Diffraction from Materials*, Academic Press, New York (1977)

12. Yee.A.F. and Smith,S.A., *Macromolecules*, **14**, 54 (1981)

13. D.J. Massa and P.P. Rusanowsky, *ACS Polymer Preprint* **17**(2), 184 (1976)

14. D. Stefan and H.L. Williams, *J. Appl. Polym. Sci.*, **18**, 1279 (1974)

15. L.J. Garfield, *J. Polym. Sci.,Part C*, **30**, 551 (1970)

16. W.J. Koros, *J.Polym. Sci. Polym. Phys. Ed.*, **23**, 1611 (1985)

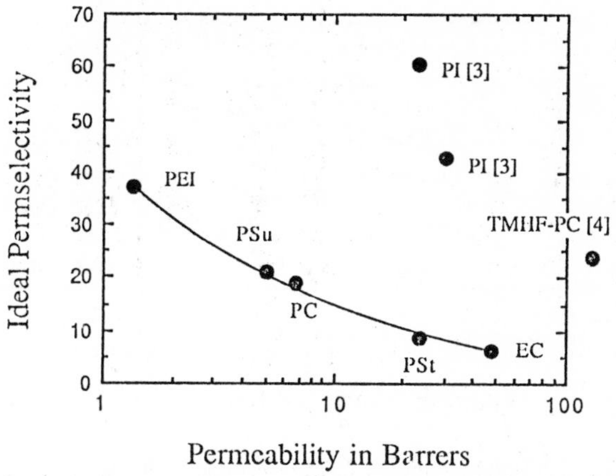

Figure 1. Permeability vs. permselectivty in glassy polymers for the CO_2/CH_4 system. The line passes through points for typical polymers. The points designated PI and TMHF-PC are data on polyimides of Kim et al. [3], and on a polycarbonate shown in Figure 2, respectively.

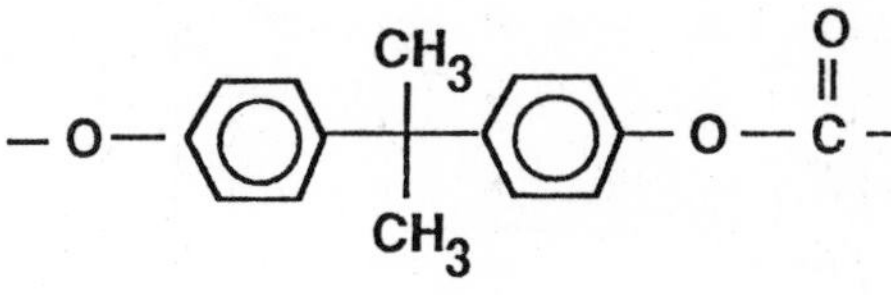

Bisphenol-A Polycarbonate (PC)

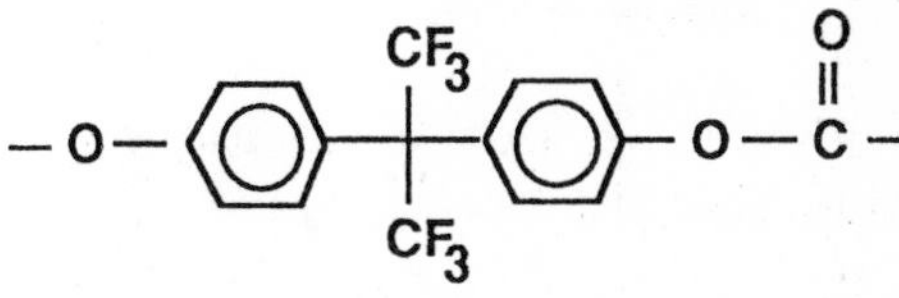

Hexafluorobisphenol-A Polycarbonate (HFPC)

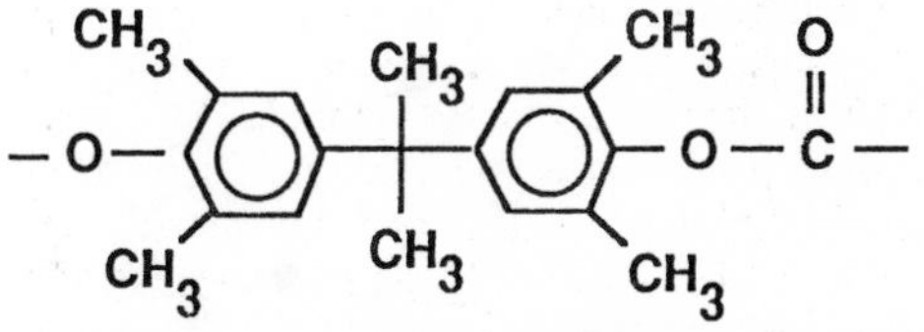

Tetramethylbisphenol-A Polycarbonate (TMPC)

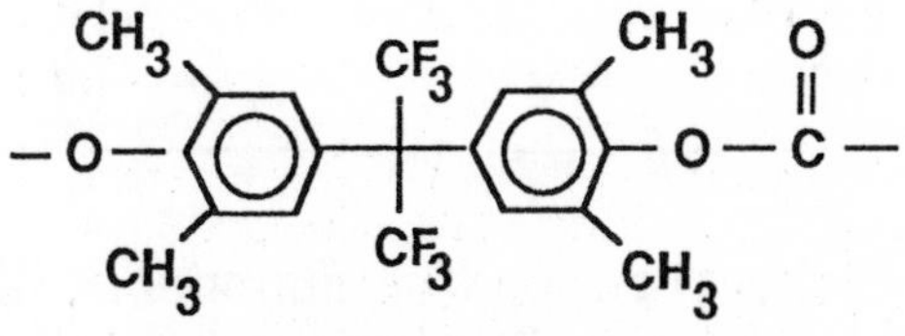

Tetramethylhexafluorobisphenol-A Polycarbonate (TMHF-PC)

Figure 2. Polycarbonate structures.

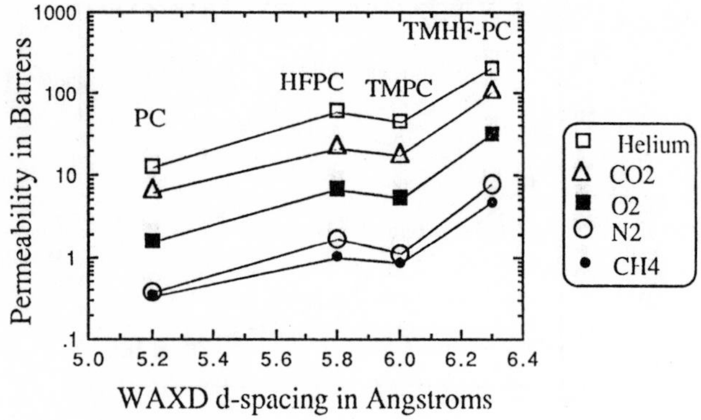

Figure 3. Permeability vs. WAXD d-spacing for various gases in the polycarbonates shown in Figure 2.

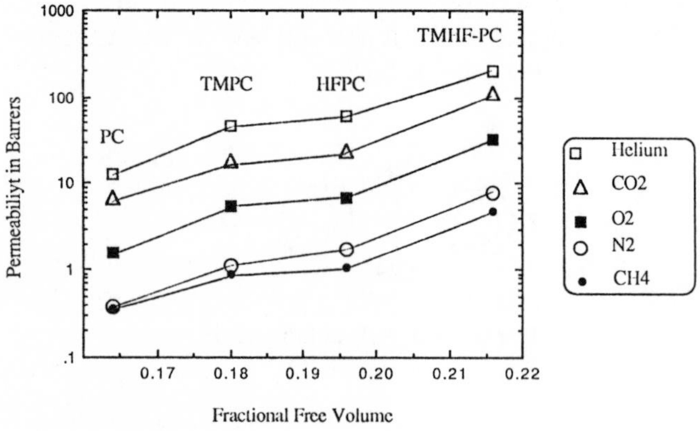

Figure 4. Permeability vs. fractional free volume for various gases in the polycarbonates shown in Figure 2.

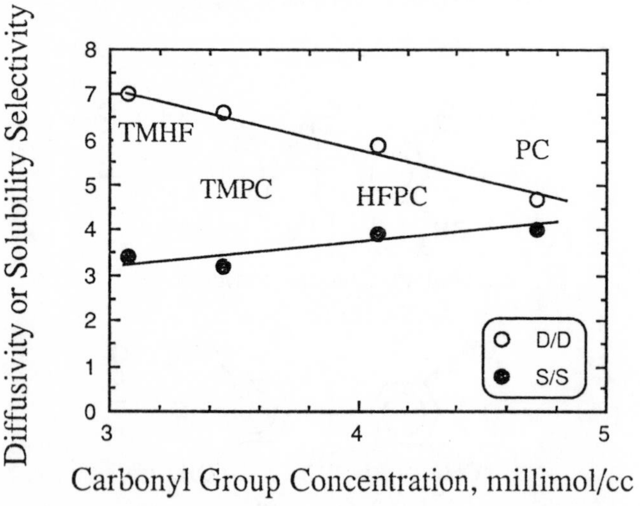

Figure 5. Diffusivity and solubility selectivity vs. carbonyl group concentration for the CO_2/CH_4 system in the polycarbonates shown in Figure 2.

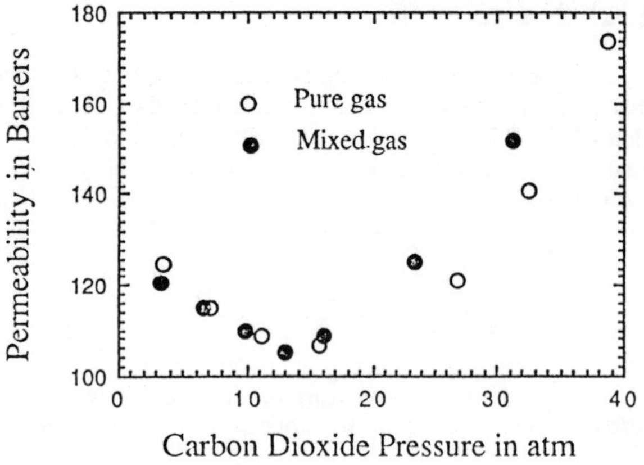

Figure 6. Pure and mixed gas permeability of carbon dioxide in TMHF-PC. The gas mixture was 50/50 mole percent CO_2/CH_4.

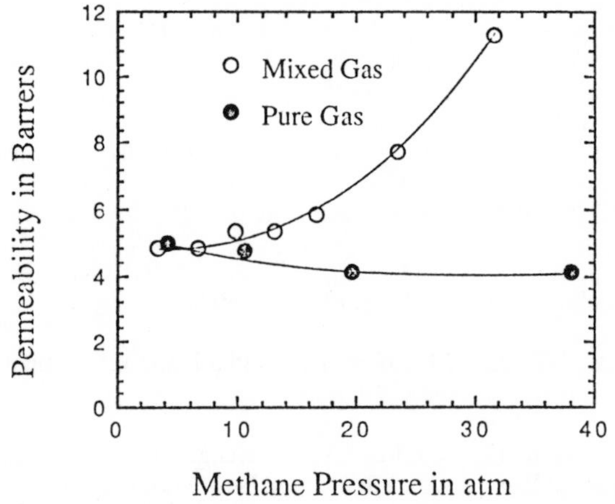

Figure 7. Pure and mixed gas permeability of methane in TMHF-PC. The gas mixture was 50/50 mole percent CO_2/CH_4.

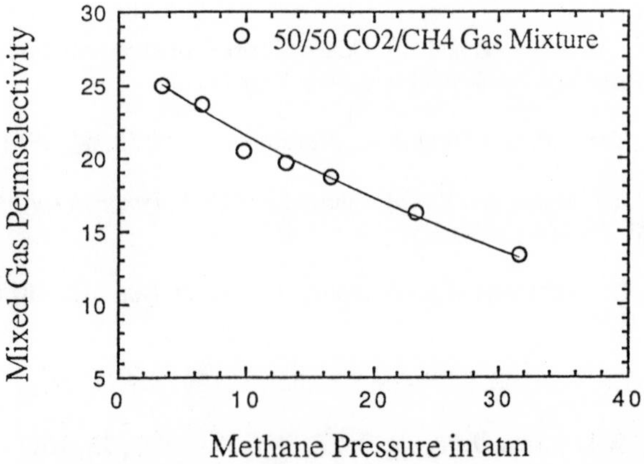

Figure 8. Mixed gas permselectivity vs. pressure in TMHF-PC. The gas mixture was 50/50 mole percent CO_2/CH_4.

SELECTIVE PERMEATION OF A GAS MIXTURE IN SURFACE MODIFIED PMMA/PVDF POLYBLEND MEMBRANES

Tahua Yang and Charles E. Rogers ■ Department of Macromolecular Science, Case Western Reserve University, Cleveland, OH 44106

Polyblends of polymethylmethacrylate (PMMA) and poly(vinylidene fluoride) (PVDF) were used as membranes to separate a gas mixture of carbon dioxide and methane. A Harshaw permeability cell with FTIR detection was used to determine the permeation properties of these membranes. Blending the two polymers improved their mechanical properties and permeability, but deteriorated the permeation selectivity due to the plasticizing action of the PVDF, which has high chain flexibility. Acid hydrolysis of the surface region changed the structure of the surface layer to a mixture of PMMA, PVDF and poly(methacrylic acid) (PMAA). Posttreatment of the partially hydrolyzed membranes by water or isopropyl alcohol modified the morphology and the packing density of the thin surface layer, which is interpreted in terms of the ionic nature of PMAA. The treatment by isopropyl alcohol after hydrolysis dramatically enhanced the permeation selectivity with little loss of permeability.

Due to economic considerations and environmental protection restrictions, membrane technology is now challenging conventional gas separation processes such as cryogenic and adsorption processes (1). As this market expands, there is increasing demand for new materials with high selectivity and high flux. To synthesize a material as a membrane for gas separation operations is a difficult task. However, there is another avenue leading to such new materials. The technology of polymer blending is a feasible way to improve the properties of existing materials to satisfy specific application requirements. The technique of polymer blending has been used extensively in industry. Those processes to improve properties of materials are much easier than new polymerizations, copolymerizations or the formation of interpenetrating matrices (2).

Surface modifications often are needed to improve properties like wettability, biocompatability, abrasion resistance and adhesion, while leaving the bulk properties unchanged (3). The surface characteristics of separation membranes have been successfully modified using plasma (4,5) and graft copolymerization procedures (6) to enhance permeation selectivity. Very often, the modified surface layer plays an important role in the permeation separation process while the bulk region still provides those good properties of the original polymer.

Correspondence should be addressed to C.E. Rogers.

Polymethylmethacrylate(PMMA) /poly (vinylidene fluoride)(PVDF) is a well-known polymer blend (7). The mechanical properties of PMMA can be considerably improved by blending it with PVDF. The ideal separation factors of various gases through these membranes have been investigated by Chiou et al. (8, 9). In this present study, we used a different method utilizing a Harshaw permeability cell with Fourier Transform Infrared detection to investigate the permeation of the gas mixture, carbon dioxide and methane, through miscible PMMA/PVDF blends and surface modified blend membranes. Use of the gas mixture as such for the gas feed allows any intereaction between the two gases to occur while they are passing through the membranes so that a measure is made of the real separation factor, defined here as alpha, equal to the ratio of the measured permeabilities of carbon dioxide/methane.

MATERIALS AND PROCEDURES

Materials

PMMA was obtained from DuPont, designated as Elvacite 2041. The second polymer, PVDF (Solef 4012), was obtained from the Soltex Corporation. Both polymers were obtained in the form of fine powders. The PMMA had a Mw of 495,000, Mn of 193,000 and Tg of 104 ^{o}C. The PVDF had a Mw of 218,000,

Mn of 81,000 and Tg of -35 °C.

Membrane Formation

PMMA and PVDF were dissolved at elevated temperature in dry dimethyl sulfoxide (DMSO). Solutions of five weight precent (Wt%) were made with different compositions of both polymers. These solutions were centrifuged and degassed before membranes were cast onto glass plates. The membranes were kept at 100 °C for at least one day to evaporate solvent and to remove residual stress and defects. The thickness of the final membranes were about ten microns.

Surface Hydrolysis Procedure

Acid hydrolysis of 30 wt% PVDF polyblend was performed by immersion under 65% sulfuric acid at room temperature. Only one side was hydrolyzed, the other side was covered by a Teflon plate (10). The membranes were treated with washes of water or isopropyl alcohol after having been hydrolyzed for various periods of time.

Differential Scanning Calorimetry

The measurements of Differential Scanning Calorimetry (DSC) were carried out in a Perkin Elmer 7 Series Thermal Analysis System. The heating rate employ for all samples was 10 degrees per minute. The instrument was first calibrated with indium.

Permeation Measurements

Figure 1 shows a schematic of the permeation apparatus which includes a temperature regulator (R) and a thermal bath (H) to maintain the temperature at 25 ±1 °C. A 50% gas mixture (A) of carbon dioxide and methane was purchased from Matheson Gas Company. Nitrogen (B) was the purging gas for the downstream chamber (upper part of permeability cell) before a measurement was started. The permeability cell (P) contains two chambers with the membrane (M) mounted between them. The feed gas mixture flows through the lower chamber. Permeation occurs through the membrane into the upper chamber. The two valves (v) of the upper chamber are closed during the course of a measurement. The whole permeability cell is installed inside a FTIR chamber and the concentration of the two gases collected in the upper chamber can be measured through IR transparent windows. No supporting films (such as paper filters) were used. A mercury bridge (K) balanced the pressures between the two chambers to prevent rupture or deformation of the membrane. The pressure of measurement was one atm. At that low pressure there is no effect of plasticization induced by the sorption of carbon dioxide.

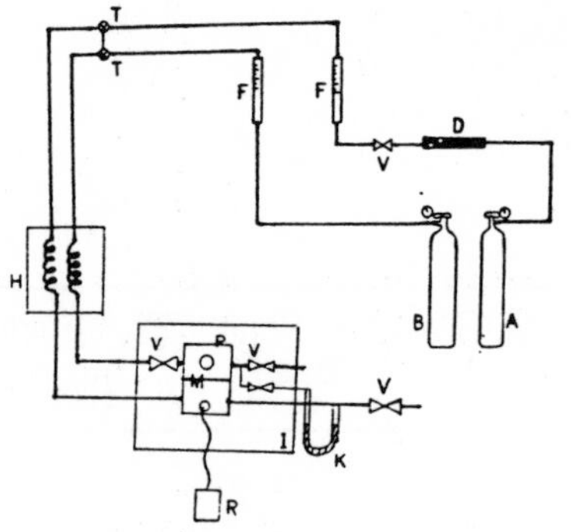

Figure 1. Schematic of apparatus for permeation measurement.

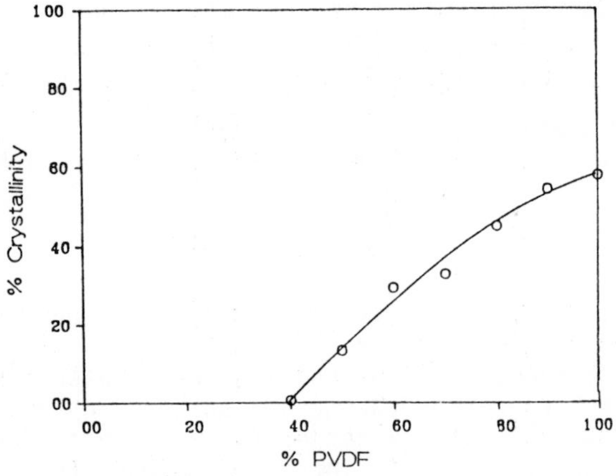

Figure 2. Crystalline contents measured by DSC.

RESULTS AND DISCUSSION

Unhydrolyzed Polymer Blends

Figure 2 shows the percentage of crystallinity, measured by DSC, in the polyblends of different compositions. Those membranes with a content of PVDF less than 40 wt% show an amorphous (miscible) conformation. When the concentration of PVDF is higher than 40 wt%, phase separation of crystalline and amorphous regions can be observed. The results are basically in agreement with Chiou's (8) observations on extruded membranes. The uneven distribution of the two polymers into amorphous and crystalline regions increases the complexity of permeation data analysis. In this study, we only focused on those miscible, amorphous polyblends with a concentration of PVDF less than 40 wt%.

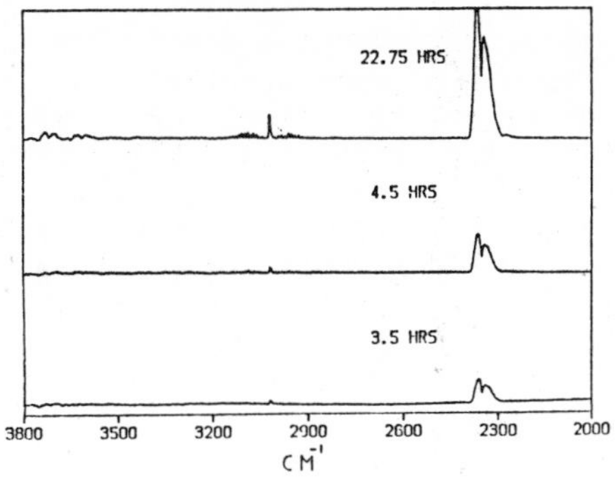

Figure 3. FTIR spectra of carbon dioxide and methane at different permeation times.

The FTIR spectra in Figure 3 show an example of permeation data for a polyblend with 20 wt% PVDF. The concentrations of gases collected in the upper chamber progressly increase with increasing time of permeation. The peak at 3016 wavenumber is that of methane with the carbon dioxide peak at 2362 wavenumber. The area of each peak is then converted to the concentration of each individual gas via a calibration curve. In this apparatus, the permeabilities of the two gases are determined simultaneously so that the actual separation factor of the gas mixture per se, can be calculated.

Figure 4 shows concentration increasing with time of permeation. The plot contains both the transient and steady state permeation regions. Apparent diffusion

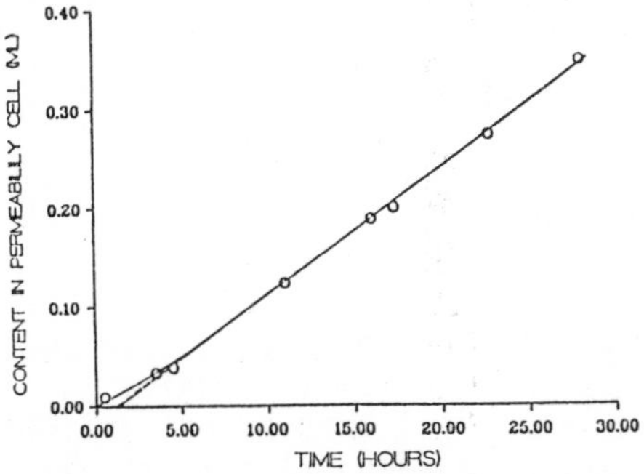

Figure 4. Concentration of methane in the upper chamber versus time of permeation. Membrane with 20 wt% PVDF.

coefficients (D) can be obtained from these data by the time lag (L) (the extrapolation of the steady state line back to zero concentration), using the usual expression: $D=l^2/6L$, where l is the membrane thickness. Steady state permeability, P, can be calculated from the slope of the straight line (11). The solubility coefficient, S, can be calculated by the definition, P = DS.

The time-scale of measuring the steady state permeability is much more than three times the time lag indicating that the reliability of the data for steady state permeability has met the criterium for attainment of the steady-state condition (11).

The relationships of permeabilities and content of PVDF in the polyblend are shown in Figures 5 and 6. The values of permeability

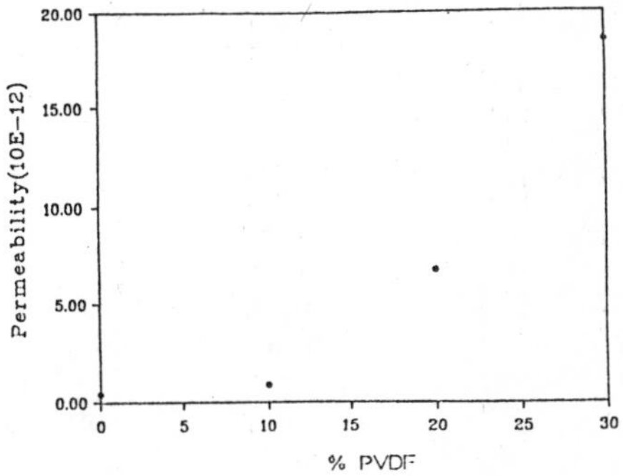

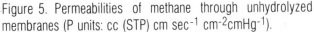

Figure 5. Permeabilities of methane through unhydrolyzed membranes (P units: cc (STP) cm sec^{-1} $cm^{-2}cmHg^{-1}$).

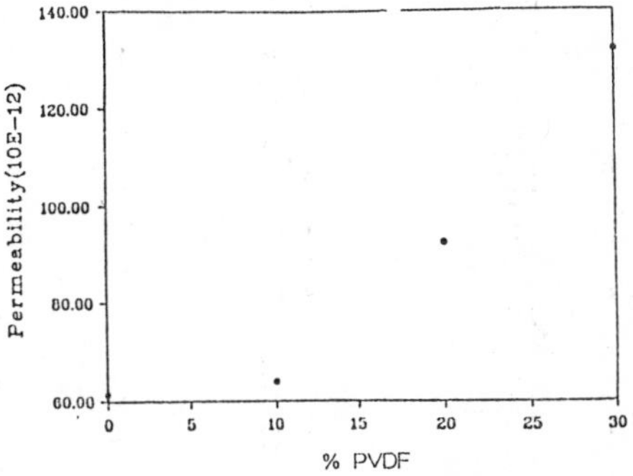

Figure 6. Permeabilities of carbon dioxide through unhydrolyzed membrane (P units: cc (STP) cm $sec^{-1}cm^{-2}cmHg^{-1}$).

of carbon dioxide and methane increase with increasing content of PVDF. This tendency was also observed and reported by Chiou et al. (8,9). Apparent diffusivities, Figure 7, calculated from time lags also show the increasing trends. The glass transition temperature of PMMA is 104 ^{o}C and that of PVDF is -35 ^{o}C. The decreasing value of glass transition temperature with increasing content of PVDF in the concentration range we studied illustrates the effect of plasticization of the blend by PVDF. PVDF has more flexible molecular chains than PMMA does. Thus, overall flexibility or mobility of molecular chains in the system is increased by increasing the content of PVDF. Gas molecules can take advantage of these factors and diffuse faster in agreement with the predictions of many free volume and other

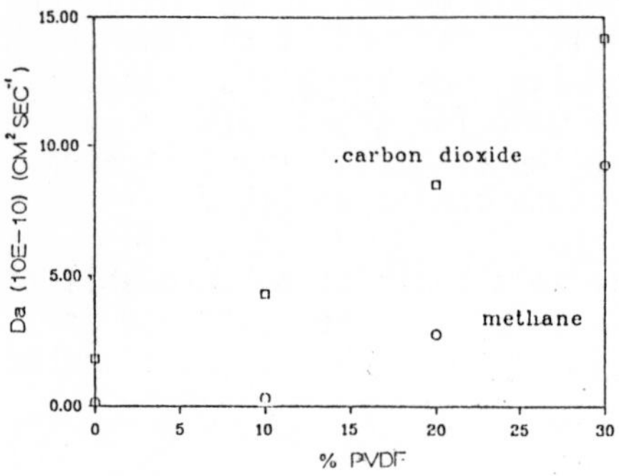

Figure 7. Apparent diffusion coefficients (cm^2sec^{-1}) of both gases with various contents of PVDF.

theories for the effects of plasticization on gas transport behavior (11). The value of alpha, the separation factor or selectivity, decreases, (Figure 8), from 139 for PMMA to 7 for 30 wt% PVDF polyblend, when both gases penetrate a membrane more easily.

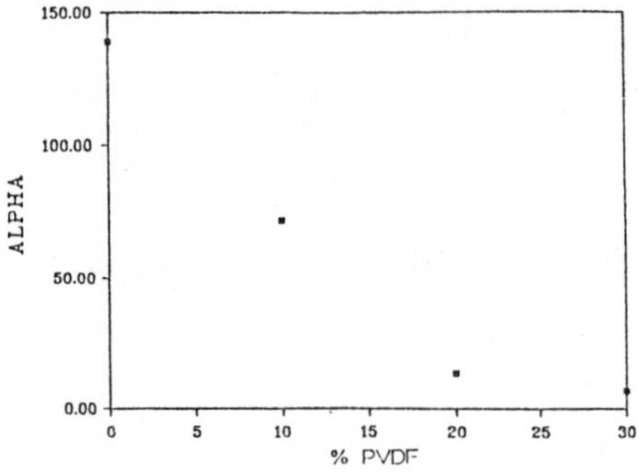

Figure 8. Separation factor variation with wt% PVDF.

These results are not unexpected. There have been several studies in which the permeabilities have been enhanced with concurrent changes in mechanical or other properties upon blending two or more polymers. It is a common enough practice to sacrifice permeability or selectivity behavior for the benefits of enhancing other properties of importance for applications other than gas permeation separation processes.

However, in our case, we want to improve gas permeation selectivity, without any or with only a small loss of of the other good properties of the polyblend. A handy and effective way to achieve this objective is to modify the surface of the polyblend membrane. This is somewhat similar to a composite membrane (12), which has a dense thin surface layer deposited upon a porous bulk layer. The porous layer offers good mechanical properties and high flux while the dense surface layer serves to separate the penetrants by some selective transport mechanism. Since the surface layer is too thin to affect other transport and mechanical properties, this general technique is quite popular for the formation of membranes for separation processes.

In our study, the requisite surface modification can be carried out without many of the experimental difficulties involved in the fabrication of composite membranes.

Surface Hydrolyzed Polyblend Membranes

Acid hydrolysis of PMMA was demonstrated by Semen and Lando (13) under the conditions of concentrated sulfuric acid at elevated temperatures. Sulfuric acid of 65% concentration was used in our studies to facilitate the surface hydrolysis. Figure 9 shows partial hydrolysis of pure PMMA film. One

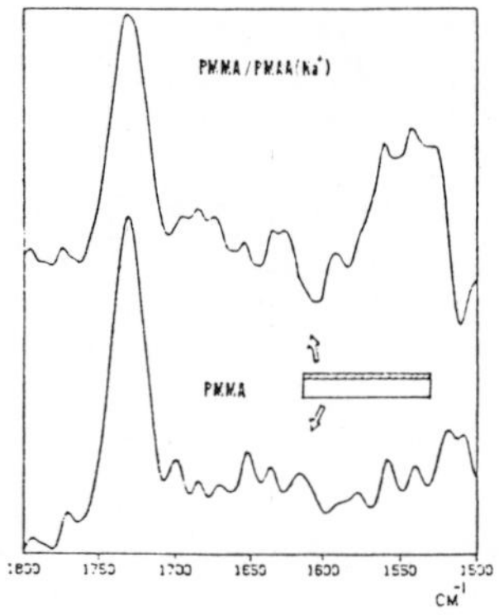

Figure 9. FTIR-ATR spectra for surface hydrolyzed PMMA film (0% PVDF).

side was reacted and neutralized by a dilute solution of sodium hydroxide. The bulk region was still transparent and the surface layer was whitened after hydrolysis. The surface layer was very carefully removed from the bulk layer and both were investigated by FTIR-ATR. The spectrum of the bulk region shows the C=O stretching band of PMMA while that of the surface layer has an extra peak at 1550 wavenumber, indicating the sodium form of the methacrylic acid group. This observation confirmed that the hydrolysis can be carried out effectively using dilute sulfuric acid rather than concentrated acid.

It should be noted, however, that hydrolysis of these same polyblend systems using concentrated sulfuric acid at 40 ^{o}C also causes pronounced changes in gas transport behavior and selectivity (4). In that case there is a marked increase in carbon dioxide permeability with only a minor increase in methane permeability. That hydrolysis procedure gives membranes with both higher selectivities and higher carbon dioxide flux. There is a degradation in other membrane properties such as cohesive strength.

In this present study, after hydrolysis using the diluted acid, the membranes were treated with either distilled, deionized water or isopropyl alcohol. The changes in permeabilities and selectivity with time of hydrolysis using a water post-treatment are shown in Figures 10 and 11. It is seen that

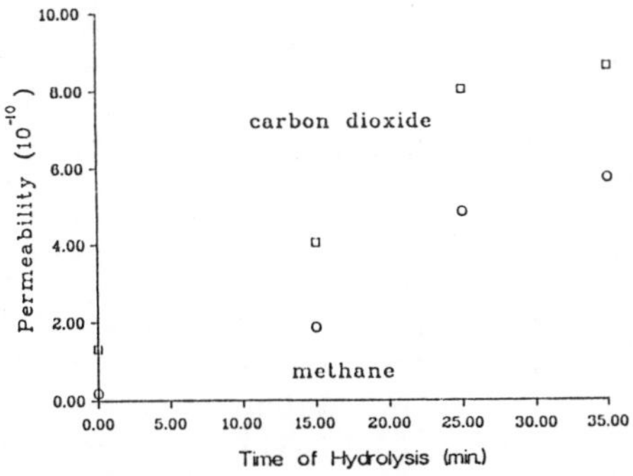

Figure 10. Permeabilities of both gases following post-treatment with water as a function of hydrolysis time.

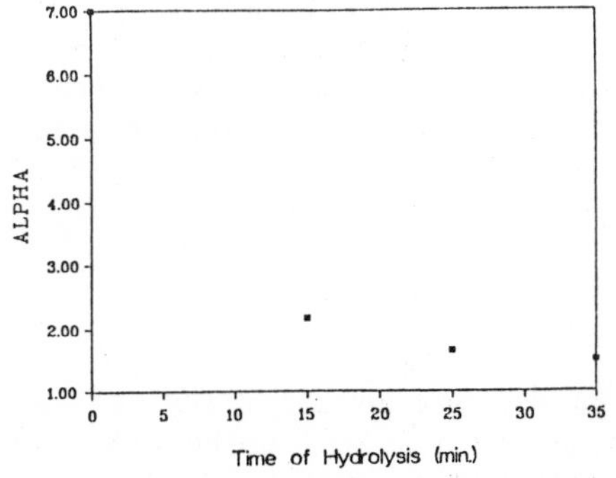

Figure 11. Separation factor variation as a function of hydrolysis time with water post-treatment.

the permeabilities of the two gases both increase with a consequent decrease in the selectivity (from 7 to 1.5 with 35 minute hydrolysis).

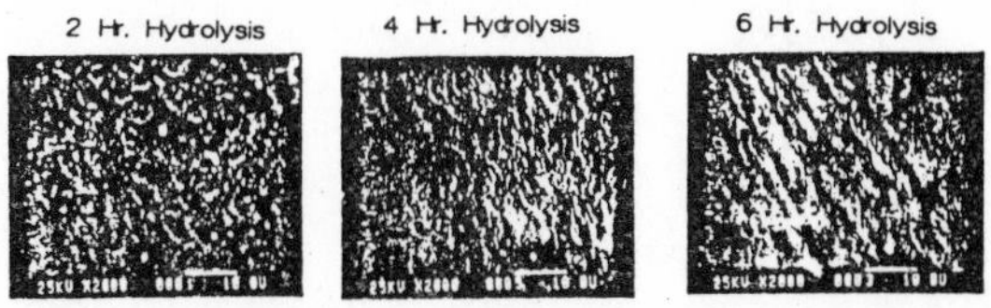

Figure 12. Surface morphologies of 30 wt% PVDF hydrolyzed membranes with water post-treatment.

These results indicate that the surface layer has a loosely packed or plasticized structure. During the course of hydrolysis, a gel-like layer was observed to form on the surface. After evaporation of the water from the post-treatment, a roughened surface was observed by electron microscopy, Figure 12. These observations are consistent with a model in which water, with its high dielectric constant (about 80), easily dissociates the proton from the methacrylic acid groups formed by the hydrolysis of the PMMA. The molecular chains then have negative charges that create repulsion forces among those chains leading to expansion of the surface layer. The permeability and selectivity behavior support this model.

The other post-treatment reagent, isopropyl alcohol, with a much lower dielectric constant (about 18), is much less effective in the dissociation of protons from acid function groups. Using this post-treatment,permeabilities of carbon dioxide and methane are slightly decreased, Figure 13, with increasing time of hydrolysis. This tendency indicates that the surface layer is now denser than the bulk. Figure 14 is a

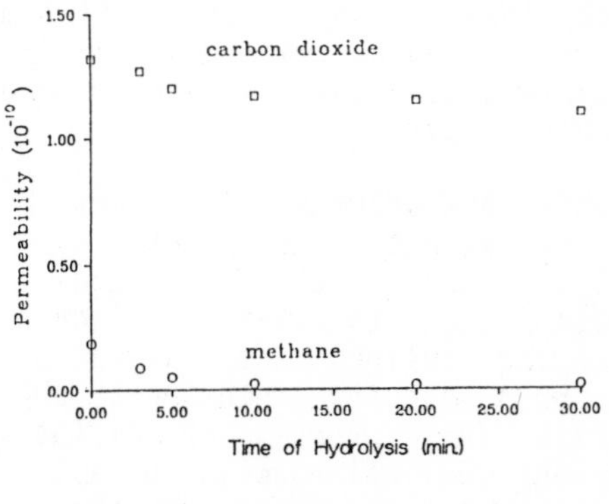

Figure 13. Permeabilities of both gases with isopropyl alcohol post-treatment.

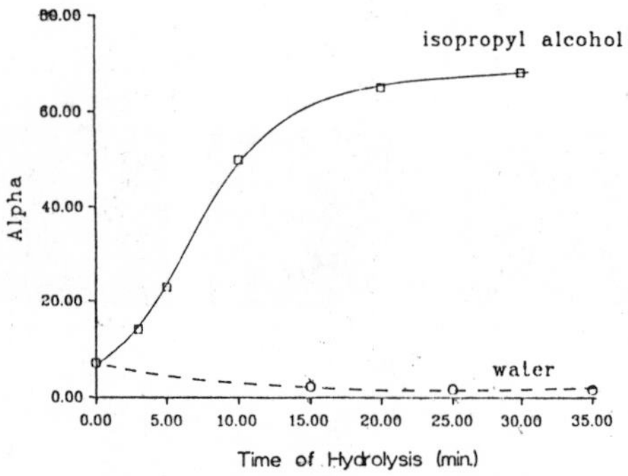

Figure 14. Comparison of separation factors with water and isopropyl alcohol post-treatment.

comparison of the permeabilities of isopropyl alcohol and water post-treatment membranes which shows the dramatic differences in their alpha values. The range of the increase in alpha for the isopropyl alcohol post-treatment membranes is significant in terms of gas separation processes.

It is to be noted that the curve for the isopropyl alcohol data in Figure 14 is somewhat S-shaped. The initial rate of change is slower than that between five to ten minutes of hydrolysis. After ten minutes the rate of change decreases again, leading to an apparent limiting value after about 30 minutes of hydrolysis. This behavior can be attributed to the nature of the sorption-reaction process during hydrolysis, the relative polymer compositions of the initial surface and bulk of the polyblend and the model of permeation in a hydrolyzed region discussed above.

We have shown in a related study (15) that the surface of these PMMA/PVDF polyblends are enriched in PVDF relative to the overall and bulk compositions. This experimentally confirmed result is expected in terms of thermodynamics which predicts that the lower energy component of a mixture, in this case PVDF, will tend to migrate to the surface to obtain an overall lower surface energy state.

This lack of PMMA in the surface regions would be expected to affect the rate of hydrolysis during the inital stages. It is estimated that the advancing front of the penetrating sulfuric acid reagent will be at about 50 Å into the membrane at about five minutes. This process of diffusion with concurrent chemical reaction usually follows kinetics which are similar to those ascribed to the Case II sorption process; a linear advance of a sorption front with time (11).

After about five minutes of hydrolysis, the advancing front reaches regions which are richer in PMMA, due first to the absence of PVDF which has migrated to the surface region and then to the expected concentration of PMMA of the original bulk compositon. Thus, the rate of hydrolysis increases by the law of mass action. As the front advances into the film, the concentration of hydrolyzed material normalizes and the rate of hydrolysis decreases to a limiting level characteristic of the reaction at that temperature.

The nature of the hydrolysis process suggests that the modified, hydrolyzed region may approximate a laminate structure over the unreacted bulk material. More realistically, there may be some gradient of reaction products in the reaction zone along the diffusion path of the acid reagent. This laminate would be expected to affect the diffusivity and, especially, the solubility of the carbon dioxide and methane penetrant molecules.

The exact details of this discrimination selectivity process are not clear at this time, but it may well involve chain flexibility or free volume considerations or some specific interaction between the hydrolyzed species, as modified by the isopropyl alcohol post-treatment, with the penetrant molecules. In any case, the result is to increase the permeation of carbon dioxide relative to that of methane, increasing the selectivity of the hydrolyzed and post-treated membrane in the favor of carbon dioxide. It is anticipated that this general mechanism can be optimized to provide a membrane system that will possess both a high selectivity and a high flux for carbon dioxide and have favorable mechanical and other properties suitable for gas separation applications.

CONCLUSIONS

The Harshaw permeability cell with FTIR detection is a useful method to investigate the transport properties of gas mixtures in polymeric membranes. The high precision of measurement allows detection of subtle changes in transport behavior with all components of the gas mixture sorbing and diffusing concurrently.

Polymer blends of PMMA/PVDF show enhanced mechanical properties and permeabilities of carbon dioxide and methane. However, the separation factor, alpha (the ratio of the permeabilities of CO_2 and CH_4 as measured) is significantly decreased from 139 to 7 as the PVDF content is increased from zero to 30% by weight. This trend is generally found when membranes are plasticized, increasing the permeation rate of all penetrant species.

Surface modification by hydrolysis of PMMA to PMAA using diluted sulfuric acid markedly changes the transport behavior. Post-treatments of the hydrolyzed samples with water further decrease alpha from 7 to 1.5 for 35 minutes of hydrolysis. However, the use of isopropyl alcohol for post-treatment increases alpha from 7 to 67 for the same time of hydrolysis.

REFERENCES

1. R. W. Spillman, Chem. Eng. Prog., Jan, 41, (1989)
2. I. Cabasso, Kirk-Othmer, Encycl. of Chem., Tech., **12**, 492, (1980)
3. ACS Symp. Ser. No. 162, D. W. Dwight, T. J. Fabish and H. P. Thomas Ed. (1981), Washington, DC.
4. N. Inagaki and H. Katsuoka, J. Mem. Sci., **34**, 297, (1987).
5. N. Inagaki, N. Kobayashi and M. Matsushima, J. Mem. Sci., **38**, 85 (1988).
6. H. Iwata and T. Matsuda, J. Mem. Sci., **38**, 185 (1988).
7. E. Roerdink and G. Challa, Polymer, **19**, 173 (1978).
8. J. S. Chiou and D. R. Paul, J. Appl. Polym. Sci., **32**, 2897 (1986).
9. J. S. Chiou and D. R. Paul, J. Appl. Polym. Sci., **32**, 4793 (1986).
10. T. Yang, Ph.D. Thesis, Department of Macromolecular Science, Case Western Reserve University, Cleveland, Ohio, 1989
11. C. E. Rogers, "Permeation of Gases and Vapours in Polymers", in Polymer Permeability, J. Comyn, Ed., Elsevier Appl. Sci. Publ., Ltd., London, 1985, Chap 2.
12. M. Niwa, H. Ohya, Y. Tanaka, N. Yoshikawa, K. Matsumoto and Y. Negishi, J. Mem. Sci., **39**, 301 (1988).
13. J. Semen and J. B. Lando, Macromolecules, **2**, 570 (1969).
14. R. Duran, L. Kim, C. E. Rogers and T. Yang, J. Membrane Sci., submitted.
15. T. Yang and C. E. Rogers, March meeting of the American Physical Society, St. Louis, MO (1989).

EFFECT OF DRYING CONDITIONS ON THE PERFORMANCE AND QUALITY OF SYNTHETIC MEMBRANES USED FOR GAS SEPARATIONS

C. Yong, A.E. Fouda and T. Matsuura ■ Division of Chemistry, National Research Council of Canada, Ottawa, K1A-OR6, Canada

Dry cellulose acetate membranes were prepared by the solvent exchange technique. Different pore sizes of the dry membranes were obtained by shrinking the water-wet reverse osmosis membranes at various temperatures, then isopropyl alcohol was used to replace the water in the wet membranes then a second solvent to replace the first one before drying.

Hexane was chosen as a second solvent due to its minimal effect on the membranes during the drying procedure. The drying procedure was modified to study the effect of the evaporation rate of the second solvent (hexane) on the produced membranes. It was found that the product membranes were superior in appearance with minimal shrinkage and very smooth surfaces.

The membranes were characterized to determine pore size and pore size distribution of the skin layer by using the pure gas permeation data. The pure gases tested were helium, carbon dioxide, and methane. The separation performance of the membranes were tested for the binary gas mixture of CO_2/CH_4 with five different compositions. The permeation and separation data are interpreted in terms of the surface force-pore flow model in which the relative contribution of each flow mechanism (Knudsen, Slip, Viscous, and surface) are presented.

INTRODUCTION

The use of synthetic membranes in gas separations continues to gain increasing importance in the field of membrane separation processes. Dry cellulose acetate membranes are usually prepared by the solvent exchange technique (1) to (6) in order to preserve the membranes structure that is critical to the separatory performance and practical use in gas separation applications.

The pore size and pore size distribution on the skin layer of the asymmetric cellulose acetate membranes control to a large extent the performance of the dry membranes in gas separation. In our previous studies (7) to (9), different pore sizes were obtained by shrinking the water-wet membranes at various temperatures, and by using various types of first and second solvents in the drying process. In the present work hexane was chosen to be the second solvent since it was found to have minimal effect on the membranes during drying. The drying procedure was also modified to study the performance of membranes dried under various vapor pressure levels of the second solvent.

This paper is considered to be an extension to our previous studies (7), (8), which aimed to establish a cause and an effect relationship between the various variables involved in the membranes formation and the performance of these membranes. It was shown (8) that a critical pore size exists on the surface of the water-wet membrane, that results in the smallest pore size on the dry membrane which means the highest separation factor. Such pore sizes become greater if the membrane was dried with a second solvent that has a higher boiling point. In this work we are concerned with the effect of the evaporation rate of the second solvent on the structure and performance of the dry membranes. The results are compared to the corresponding results obtained from the membranes prepared under the same conditions but dried in the air.

The domination of various gas transport components (such as Knudsen and surface flow components) was investigated (8) in relation to the combination of first and second solvents and the pore sizes so produced by these combination. In this study we will show the relative contribution of each component to the total permeation rate for the different gases.

THEORETICAL

The Surface Force - Pore Flow model assumes that the effective membrane surface layer of an asymmetric porous cellulose acetate membrane is composed of a bundle of capillary tubes. Only the pores in the skin layer are active in the separation process. The transport through the membranes can then be described with the appropriate transport equations for an individual cylindrical pore having an average radius and length, and summed over all the pores on the surface of the membrane. For that reason it is required to know the pore size distribution e.g. the log-normal distribution as expressed by the following equation :

$$N(R) = \frac{N_t}{\sqrt{2\pi}\,\ell n\,\sigma_g} \exp -\left[\frac{1}{2}\left(\frac{\ell n\,R - \ell n\,\overline{R}}{\ell n\,\sigma_g}\right)^2\right] \quad (1)$$

For each pore, the flow of gas can be described by one of the three mechanisms, namely (1) Knudsen flow; (2) slip flow; (3) viscous flow.

The discrimination between each mechanism is dependent on the relative magnitude of the pore radius to the mean free path. In this work we adopted the limitation of Liepmann (10) of $R/\lambda < 0.05$ for Knudsen flow, while viscous flow would occur for $R/\lambda > 50$ according to Stahl (11). Between the values of $R/\lambda > 0.05$ and < 50.0, slip flow will be considered the appropriate mechanism. For a wide pore size distribution one can expect the three mechanisms to occur simultaneously, but to a different extent according to the operating conditions of pressure and temperature and the type of gas under consideration.

The mathematical derivation of the pore flow equations can be found in reference (9), which gives an expression for the total pore flow of the gas as follows:

$$Q_g = \frac{N_t\,\Delta P}{\delta}\,[G_1I_1 + G_2I_2 + G_3I_3] \quad (2)$$

where G_1, G_2, G_3 are constants depending on the physiochemical properties of gases, and the integrals I_1, I_2, and I_3 are numerical values of integrals dependent on the porous structure and defined in detail in reference (12).

Rangarajan et al (13) showed that the transport of gas molecules under the influence of gas polymer interaction can be represented by the equation :

$$Q_s = A_2\,\frac{I_4}{I_5}\,\overline{P}\Delta P \quad (3)$$

where I_4 and I_5 are also numerical values of integrals dependent on the porous structure and defined in reference (14). The total gas flow can then be described by the following equation :

$$Q_t = Q_g + Q_s \quad (4)$$

By defining the gas permeability coefficient, A_G, as the amount of gas permeating per second per unit area per unit pressure difference, and denoting the membrane area by S, we get the following expression :

$$A_G = \frac{Q_t}{S.\Delta P} = \left(\frac{N_t}{S.\delta}\right)(G_1I_1 + G_2I_2 + G_3I_3) + \frac{A_2I_4}{S\,I_5}\,\overline{P} \quad (5)$$

$$A_G = A_1(G_1I_1 + G_2I_2 + G_3I_3) + A_2\,\frac{I_4}{I_5}\,\overline{P} \quad (6)$$

Equation 6 represents the basic relationship between the permeability coefficient A_G and the average pressure $\overline{P}$ across the membrane. To be able to use Equation 6, four parameters have to be evaluated from the permeation data obtained under different operating pressures; A_1 is related to the pore structure, A_2 is related to the surface transport and $\overline{R}$ and σ_g of the log-normal distribution which are needed to calculate the integrals I_1, I_2, I_3, I_4 and I_5.

Prediction of CO_2/ CH_4 Gas Mixture Permeation Rate

The total permeation rate for the binary gas mixture can be calculated from the individual fluxes of the components as follows :

$$[PR] = J_1 + J_2 \quad (7)$$

The individual fluxes of the mixture components can be written as :

$$J_i = (A_G)_i\,(P_2\,x_{i2} - P_3\,x_{i3}) \quad (8)$$

where x_{i3} (i=1,2) is the mole fraction of gases 1 and 2 in the permeate and can be defined as follows :

$$x_{13} = \frac{J_1}{J_1 + J_2}$$

$$\text{or} \quad x_{13}(J_1 + J_2) - J_1 = 0 \qquad (9)$$

$$x_{13} + x_{23} = 1 \qquad (10)$$

The prediction of total permeation rate [PR] for the gas mixture can be done as follows:

(1) The permeation data are collected for each component of the gas mixture at different pressure gradients.

(2) The experimental permeability coefficient, A_G is calculated vs. the average pressure $\bar{P}$ across the membrane.

(3) The optimum value of the characterization parameters were calculated using either the Simplex method (14) ,or using a grid search method. In the later method, the parameters $\bar{R}$ and σ_g were assumed which determined the range of pore radius values (R_{min} and R_{max}).

(4) The integrals I1, I2, I3, I4 and I5 were calculated and substituted into Equation 6.

(5) A non-linear regression computer routine can then be used to evaluate the optimum values of the remaining parameters A1 and A2.

(6) For each combination of $\bar{R}$ and σ_g, the sum of squared residuals is calculated according to the equation

$$SS_R = \sum_{i=1}^{n} (y_i - y_i')^2 \qquad (11)$$

(7) Steps (3) to (6) are repeated at all the grid points and the parameters corresponding to the minimum value of SS_R would represent the characterization parameters of the membrane under consideration.

(8) The characterization parameters evaluated for both components (CO_2 and CH_4) are substituted into Equation 8 to determine the individual fluxes J_1 and J_2 in terms of x_{13}.

(9) The cubic Equation 9 is solved to calculate x_{13} and its value is substituted into Equation 10 to calculate x_{23}.

(10) The individual fluxes of the mixture components can then be calculated from Equation 8 and substituted into Equation 7 to calculate the total permeation rate [PR].

EXPERIMENTAL

Figure 1 is the schematic diagram of the procedure to prepare the dry asymmetric cellulose acetate membranes. The membranes were cast from a solution having the following composition (wt%): cellulose acetate (Eastman 398-3), 17%; acetone, 69.2%; magnesium perchlorate, 1.45%; and water, 12.35%. The temperature of the casting solution was kept at 10°C, and the temperature of the surrounding atmosphere was kept at 30°C, and its relative humidity was controlled at 65%. The solvent was evaporated from the cast sheet for 60s, then the sheet was transferred to a gelation bath of ice-cold water for at least one hour.

Each sheet was cut into several coupons and identified to the original sheet, and then all the membrane coupons from all the sheets were divided into several groups for heat treatment. The membranes were heat-treated (shrunk) in hot water bath for a period of 10 minutes at a controlled temperature ranging from 50°C to 90°C. Every membrane was then identified with the shrinkage temperature.

The membranes were then dried using the solvent exchange technique which is described in detail in reference (16). In this method, the water in the wet membrane was replaced by a water miscible solvent which was isopropyl alcohol. The objective of the first solvent exchange is to prevent a simple evaporation of water which has a relatively high surface tension and can cause the collapse of the pore structure during its retreat.

The first solvent was then replaced by a second solvent which included in our previous studies (7) to (9) such solvents as carbon disulfide, isopropyl ether, triethyl amine and hexane. In the present work, we limited the second solvent to hexane in order to eliminate the effect of the second solvent type on the performance of the membranes.

One of the main objectives of the present work was to study the effect of the vapor pressure of second solvent, during the drying process, on the structure and performance of the dry membranes. For that reason the drying of the membranes was done at different levels of hexane vapor pressure. This was achieved by saturating the drying atmosphere (the desiccator) from a container of a liquid mixture of hexane/cyclohexanol with any of the following hexane compositions: 25%, 50%, 75%, or 100% hexane.

The dried flat membranes were mounted in the testing cells shown in Figure 2, and flushed with the feed gas. The upstream pressure was varied in the range of 400 to 2400 kPa. The permeate flow rates were measured by soap bubble meters. The permeation data for the pure gases of helium, CO_2, and CH_4 were collected to characterize the membranes and determine their selectivity from the pure gases' permeation ratios. The separation performance of each membrane was tested for the mixture of CO_2/CH_4 with five different feed compositions mainly 10%, 20%, 50%, 80%, and 90% CO_2. All the experiments were conducted at room temperature, and the compositions of the permeate from the gas mixture experiments were measured by gas chromatography using Spectra-Physics gas chromatograph, model SP-7100, equipped with a Porapak Q column.

RESULTS AND DISCUSSION

In the present study, most of the membranes were dried in the final stage of preparation under different vapor pressures of hexane in the drying atmosphere. The immediate result of this modification in the drying process was to produce membranes with much superior surface characteristics and minimal shrinkage to the water-wet membranes. The surfaces were relatively smooth, perhaps due to slower evaporation rate of the second solvent which led to less collapse of the pore structure and minimized the surface corrugations which are produced sometimes when the membrane is dried in the air.

Pure Gas Permeation

As mentioned in a previous section, the transport of gases through the membrane pores is due to one of three mechanisms, Knudsen, slip or viscous flow. The interaction between the membrane material and the membrane causes an additional transport which we called surface flow, and is highly dependent on the gas and the membrane material. To examine the contributions of each of these four components to the total permeation of the pure gases, the relative percentages of the calculated gas transport by each of the four mechanisms (Knudsen, slip, viscous and surface), were plotted vs. upstream pressure in Figures 3 to 6.

Figure 3 shows that for helium, the slip flow mechanism is dominating the the surface flow contribution is minimum. This would be expected for helium since it is an inert gas and would have a weak interaction with the membrane material. On the contrary, the surface flow mechanism is expected to dominate for carbon dioxide gas since it has a very strong affinity to the cellulose acetate material; as is clearly shown in Figure 3.

When the water-wet membranes are shrunk at 90°C and dried in the air, much smaller pores are produced. In such case as seen in Figure 4, the Knudsen flow contribution is more visible at higher pressures more than in Figure 3, in which the membrane was originally shrunk at 60°C. In general, since the Knudsen flow limit is 0.05 λ, it will have more contribution in the smaller size pores which can be obtained by shrinking the membranes at higher temperatures.

Helium data shown in Figures 3 and 4 are plotted in Figures 5 and 6, with methane data instead of carbon dioxide. In case of methane gas, the surface flow contribution would be minimal, and most of the contribution is due to Knudsen or slip flow with variable percentage according to the pore sizes and the mean free path value at the operating pressure.

Effect of Different Preparation Parameters on Pure Gases

Shrinkage Temperature. The shrinkage temperature can be considered as one of the important factors to control the pore size on the skin layer of the asymmetric membranes. The higher the shrinkage temperature, the smaller the pore sizes, which means lower permeability and usually leads to higher separation. This fact is demonstrated in Figure 7 when the CO_2 permeation rates were plotted vs. the upstream pressure with the shrinkage temperature as a parameter. The top plot in Figure 7 contains the permeation data obtained from the second characterization set for the membrane. Both plots confirm the trend of higher permeability at lower shrinkage temperatures, as well as the increase with the inlet pressure.

Vapor Pressure. Figure 8 shows helium permeation rates for a group of membranes which were dried at 70°C, but dried at various levels of hexane vapor pressure. The figure contains in the bottom plot the permeation rates obtained when the membranes were initially mounted. It also contains in the top plot the data obtained after operating the same membranes for some period of time using both pure gases and gas mixtures.

Figure 8 shows that the permeation rates in the top plot are higher than the ones in the bottom plot. The observation can be explained by referring to our previous simulation work (16). We showed that the pore size distribution can be altered by applying some physical changes such as shrinkage at high temperatures. In a similar way, operating the membranes under high pressures for a period of time may also cause changes to the pore structure which can lead to the disappearance of smaller pores with attendant enlargement of some of the larger pores nearby. Such change would cause higher permeation rates and possibly lower separation factors.

It is also shown in this figure that the permeation rates in the membrane dried in air are generally lower than the rates obtained from membranes dried under saturated atmosphere of hexane/cyclohexanol mixture. Figure 9 shows very similar data for methane, but the permeation rates for CH_4 are generally much lower than the data for helium.

The shrinkage temperature in both Figures 8 and 9 was 70°C, which produced relatively large pores that would tend to collapse when the membrane is dried in the air. However, if the membrane was shrunk at higher temperature as shown in Figure 10 (80°C), the air dried membrane has relatively higher permeation rates, indicating that the smaller pores produced were less affected by the drying atmosphere.

The separation property for the membrane mounted in cell (2-6) is shown in Figure 9 as a plot of CO_2/CH_4 permeation ratio vs. the upstream pressure. Although the membrane in that cell has the lowest permeation rate in Figure 8, it has the highest CO_2/CH_4 permeation ratio in Figure 11. This is in general a trade off between permeability and separation characteristics of the membranes. For example, the process of drying the membranes under saturated atmosphere of the second solvent is quite attractive to produce high quality membranes, but with much lower separation factors. This is not the case for the pressure effect on the membrane performance, as Figure 11 shows a very slight effect of the pressure. This means that increasing the operating pressure would increase the productivity without sacrificing the separation capability of the membranes.

Gas mixture experiments and prediction of permeation rates. Table 1 shows the preparation conditions under which the membranes used in gas mixture experiments were prepared. Each group shown in the first column, contains several membranes prepared under the same conditions except one parameter that changed. For example all the membranes in group (1) were shrunk at 60°C but each one was dried under a different level of hexane vapor pressure, in the meantime all the membranes in group (4) were dried in the air, but each membrane was shrunk in the water-wet state at a different temperature.

Figures 12 to 14 show the effect of upstream pressure on the outlet CO_2 composition for all the five compositions that were investigated. In all the figures, a minimum change is noticed in the outlet composition. In most cases, good separation was obtained, e.g. in Figure 14, for an inlet composition of 20% CO_2, an average outlet composition of 78% CO_2 was obtained. This gives an actual separation factor of 16 according to the equation:

$$\alpha_{AB} = \frac{y_A/y_B}{x_A/x_B} \tag{12}$$

where y_A and y_B are the mole fractions of A and B in the permeate, and x_A and x_B are the mole fractions in the feed.

Surface Force - Pore Flow Model Prediction

In our previous studies (7) to (9), we compared two procedures to predict the total gas mixture permeation rates. The first procedure followed by Mazid et al. (12), which relies on the permeation data of a reference gas, was found to be unsatisfactory. It resulted in a large scatter in the data, and the calculated permeation rates were either overpredicted or underpredicted. We also showed a clear discrimination between various feed compositions. Another successful approach was introduced, in which the characterization parameters (R, σ_g, A_1 and A_2) evaluated for both components (CO_2 and CH_4) were substituted in Equation 8 to calculate the component flux.

The total permeation rates for the gas mixtures, calculated using this approach are shown in Figures 15 to 17 for three membranes prepared under the same conditions, but dried under different drying conditions. Figure 15 shows the results for the membrane that was dried in air, while Figures 16 and 17 show the results for membranes dried in saturated atmosphere of 25%, 75% hexane respectively. The bottom plots in Figures 15 and 16 which are based on the initial characterization parameters show remarkable difference between the calculated and the experimental data, when compared with the predicted data obtained from the second set of characterization parameters.

The correlation between the time of characterization and the accuracy of predicting the total permeation rates were further confirmed by characterizing some of the membranes at four different times during the experiments. A sample of these results are shown in Figures 17 and 18, which show the prediction of the same experimental data at each time. It is noticed that the improvement is quite clear between the first and second set, but no improvement in the prediction is noticed between second, third or fourth time.

Although that was a general trend of improved prediction with consecutive characterization, sometimes more stable membranes were produced and very similar predictions occurred irrelevant of the time of characterization. A typical membrane is shown in Figure 19, where the predicted results remained unchanged each time.

In general the Surface Force-Pore Flow model is quite adequate to describe the transport of both pure gases and gas mixtures. However the parameters needed to predict the transport of the gas mixtures and which are evaluated from the pure gas permeation data, should be checked after the membranes have been subjected to the flow of gases under variable pressure conditions for a significant period of time.

CONCLUSION

By controlling the vapor pressure during the final stage of drying the membranes, superior membranes are obtained with very smooth surfaces and minimum shrinkage. However, there were no remarkable improvement in the performance of these membranes to separate the CO_2/CH_4 gas mixture. The membranes dried in the air had better separation characteristics, but the permeation rates were generally lower than the membranes dried under higher vapor pressure of the second solvent.

There was no apparent effect for the uptstream pressure on the separation factors, but the permeation rates increased with the pressure.

The characterization parameters should be evaluated after the membranes have reached a steady state, and not immediately after mounting them in the cells.

NOMENCLATURE

A_1 = constant for a given membrane related to the porous structure.

A_2 = constant related to surface transport, $kmol/(m^3.s.Pa^2)$

A_G = gas permeability coefficient, $kmol/(m^2.s.Pa)$

G_1, G_2, G_3 = constants depending on the physiochemical properties of gases.

I_1, I_2, I_3, I_4, I_5 = numerical values of integrals dependent on the porous structure.

J_i = flux of gas i, $kmol/m^2s$

N(R) = number of pores having a radius R, m

N_t = total number of pores, having a radii from R_{min} to R_{max}

n = number of data points

P_2 = pressure (absolute) on the high pressure side of the membrane, Pa

P_3 = pressure (absolute) on the low pressure side of the membrane, Pa

ΔP = pressure differential across the membrane, Pa

$\overline{P}$ = mean pressure across the membrane, Pa

[PR] = permeation rate, $kmol/s.m^2$

Q_g, Q_s, Q_t = quantity of gas transported in the gas phase (total of Knudsen, slip and viscous flows), by surface flow mechanism and total quantity of gas transported, respectively, kmol/s

R = pore radius, m

$\bar{R}$ = mean pore radius, m

R_{max} = pore radius of the largest pore, m

R_{min} = pore radius of the smalles pore, m

S = membrane area, m^2

SS_R = sum of squared residuals defined by equation (11)

x_{i_2} = mole fraction of gas i on the high pressure side.

x_{i_3} = mole fraction of gas i on the permeate side

y_i = experimental value of gas permeability coefficient, kmol/m.s.Pa

y_i = predicted value of gas permeability coefficient, kmol/m.s.Pa

Greek Letters

δ = equivalent thickness of the membrane, m

λ = mean free path of gases, m

σ_g = geometric standard deviation for the log-normal pore size distribution

Subscripts

1 = carbon dioxide

2 = methane

LITERATURE CITED

1. MacDonald, W. and C. Pan, "Method of Drying Water-Wet Membranes", U.S. Patent 3,842,515, (October 22, 1974).

2. Manos, P., "Membrane Drying Process", U.S. Patent 4,080,743, (March 28, 1978).

3. Manos, P., "Gas Separation Membrane Drying with Water Replacement Liquid", U.S. Patent 4,080,744, (March 28, 1978).

4. Manos, P., "Solvent Drying of Cellulose Ester Membranes", U.S. Patent 4,068,387, (Jan. 17, 1978).

5. Manos, P., "Solvent Exchange Drying of Membranes for Gas Separation", U.S. Patent 4,120,098 (1978).

6. Minhas, B.S., T. Matsuura and S. Sourirajan, "Solvent-Exchange Drying of Cellulose Acetate Membranes for Separation of Hydrogen-Methane Gas Mixtures", ACS Symposium Series, S. Sourirajan and T. Matsuura, Eds., Vol. 281, Paper 33, pp. 451-466 (1984).

7. Lui, A., F.D.F. Talbot, A. Fouda, T. Matsuura and S. Sourirajan, "Studies on the Solvent Exchange Technique for Making Dry Cellulose Acetate Membranes for Separation of Gaseous Mixtures", J. Appl. Polym. Sci., 36, 1809-1820 (1988).

8. Lui, A., F.D.F. Talbot, S. Sourirajan, A. Fouda and T. Matsuura, "Studies on Gas Transport Through Dry Cellulose Acetate Membranes Prepared by Solvent Exchange Technique", Separation Science and Technology, 23(12&13), pp. 1839-1852 (1988).

9. Fouda, A.E., T. Matsuura and A. Lui, "Permeation of Gas Mixtures in Cellulose Acetate Membranes - Practical Approach to Predict the Permeation Rate CO_2/CH_4 Mixture", Separation Science and Technology, 23(12&13), pp. 2175-2190 (1988).

10. Liepman, H.W., J. Fluids Mechanics, 10, 65 (1961).

11. Stahl, D.E., "Transition Range Flow Through Microporous Vycor", Ph.D. Thesis, Chem. Eng. Dept., The University of Iowa, Iowa city, IA (1971).

12. Mazid, M.A., R. Rangarajan, T. Matsuura and S. Sourirajan, "Separation of Hydrogen-Methane Gas Mixture by Permeation Under Pressure Through Porous Cellulose Acetate Membranes", Ind. Eng. Chem. Process Des. Dev., 24, pp. 907-913 (1985).

13. Rangarajan, R., M.A. Mazid, T. Matsuura and S. Sourirajan, "Permeation of Pure Gases Under Pressure Through Asymmetric Porous Membranes. Membrane Characterization and Prediction of Performance", Ind. Eng. Chem. Process Des. Dev., 23, pp. 79-87 (1984).

14. Sourirajan, S. and T. Matsuura, "Reverse Osmosis/Ultrafiltration Process Principle", National Research Council of Canada, Ottawa, page 5 (1985).

15. Tremblay, A.Y., A.E. Fouda, A. Lui, T. Matsuura and S. Sourirajan, "The Use of Simplex Method to Characterize Dry Cellulose Acetate Membranes for Gas Separation", Can. J. Chem. Eng., 66, pp. 1027-1030, Dec. (1988).

16. Fouda, A.E., O. Kutowy and C.E. Capes, "Simulation of the Effect of Membrane Shrinkage on Separation Behaviour", J. Separ. Proc. Technol., 8, pp. 1-10 (1987).

Table 1. Gas mixture experiments using cellulose acetate membranes.

Cell No.	Cell ID	Temp. ° C	Solvents (First/Second)	Drying Conditions
(1-2)	(M-1-12)	60	(ISO/HEX)	(100%HEX-CYCLO)
(1-3)	(M-1-06)	60	(ISO/HEX)	(75%HEX-CYCLO)
(1-4)	(M-2-02)	60	(ISO/HEX)	(50%HEX-CYCLO)
(1-5)	(M-1-05)	60	(ISO/HEX)	(25%HEX-CYCLO)
(1-6)	(M-2-07)	60	(ISO/HEX)	(DRIRITE)
(2-2)	(M-1-9)	70	(ISO-HEX)	(100%HEX-CYCRO)
(2-3)	(M-1-8)	70	(ISO-HEX)	(75%HEX-CYCRO)
(2-4)	(M-1-10)	70	(ISO-HEX)	(50%HEX-CYCRO)
(2-5)	(M-2-4)	70	(ISO-HEX)	(25%HEX-CYCRO)
(2-6)	(M-2-6)	70	(ISO-HEX)	(DRIERITE)
(4-4)	(M-5-6)	50	(ISO/HEX)	(DRIERITE)
(4-5)	(M-5-5)	60	(ISO/HEX)	(DRIERITE)
(4-6)	(M-5-15)	70	(ISO/HEX)	(DRIERITE)
(4-7)	(M-5-7)	80	(ISO/HEX)	(DRIERITE)
(4-8)	(M-5-8)	90	(ISO/HEX)	(DRIERITE)
(5-2)	(M-7-9)	90	(ISO/HEX)	(DRIERITE)
(5-3)	(M-3-9)	80	(ISO/HEX)	(DRIERITE)
(5-4)	(M-3-14)	70	(ISO/HEX)	(DRIERITE)
(5-6)	(M-3-15)	50	(ISO/HEX)	(DRIERITE)
(5-7)	(M-6-19)	25	(ISO/HEX)	(DRIERITE)
(6-2)	(M-5-1)	90	(ISO/HEX)	(25%HEX-CYCLO)
(6-3)	(M-5-2)	70	(ISO/HEX)	(25%HEX-CYCLO)
(6-4)	(M-3-11)	90	(ISO/HEX)	(50%HEX-CYCLO)
(6-5)	(M-2-11)	60	(ISO/HEX)	(50%HEX-CYCLO)
(6-6)	(M-2-9)	80	(ISO/HEX)	(75%HEX-CYCLO)
(6-8)	(M-5-11)	50	(ISO/HEX)	(100%HEX-CYCLO)

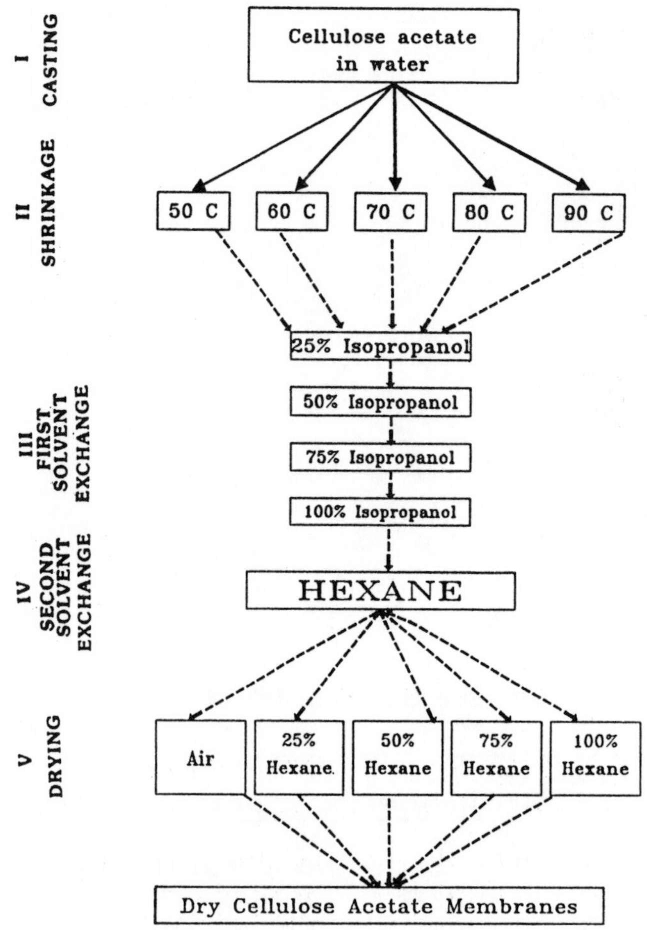

Figure 1. Schematic diagram for the preparation of dry cellulose acetate membranes by solvent exchange.

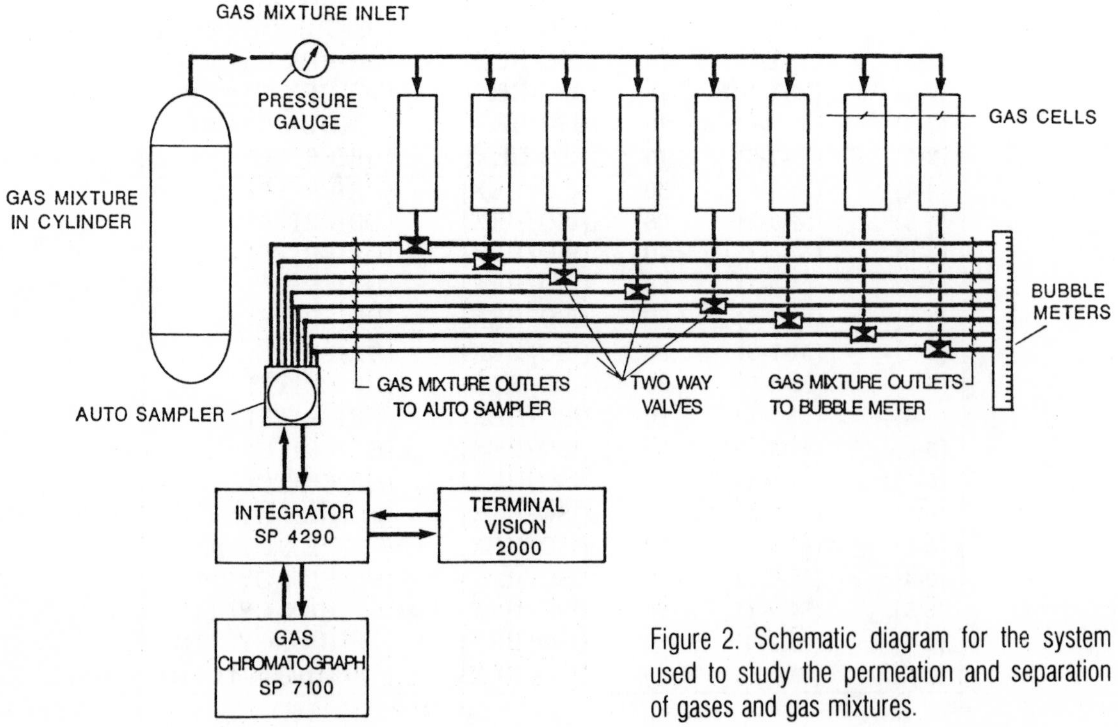

Figure 2. Schematic diagram for the system used to study the permeation and separation of gases and gas mixtures.

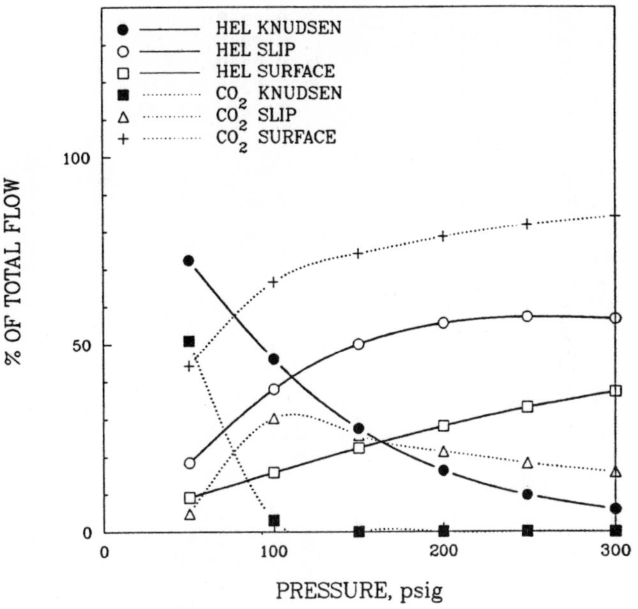

Figure 3. Contribution of flow components to the total flow of helium and CO_2 for the membrane mounted in cell (1-3).

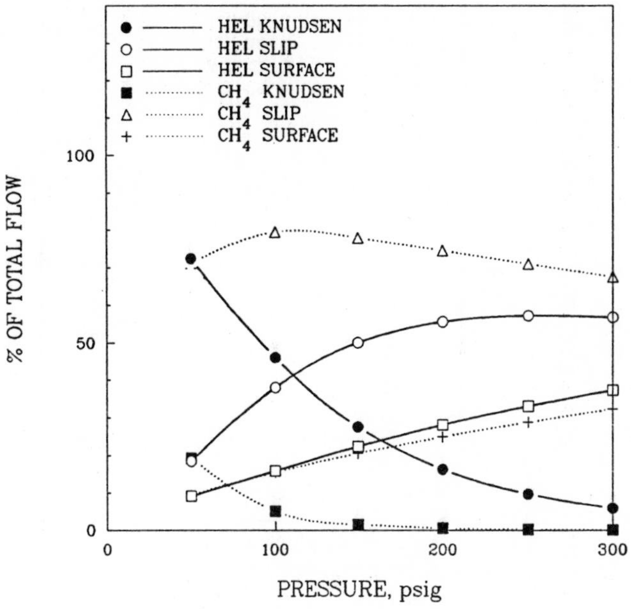

Figure 5. Contribution of flow components to the total flow of helium and CH_4 for the membrane mounted in cell (1-3).

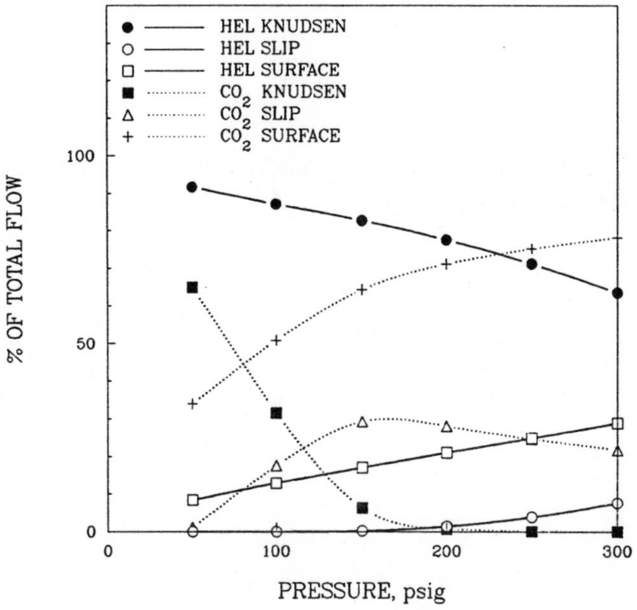

Figure 4. Contribution of flow components to the total flow of helium and CO_2 for the membrane mounted in cell (4-8).

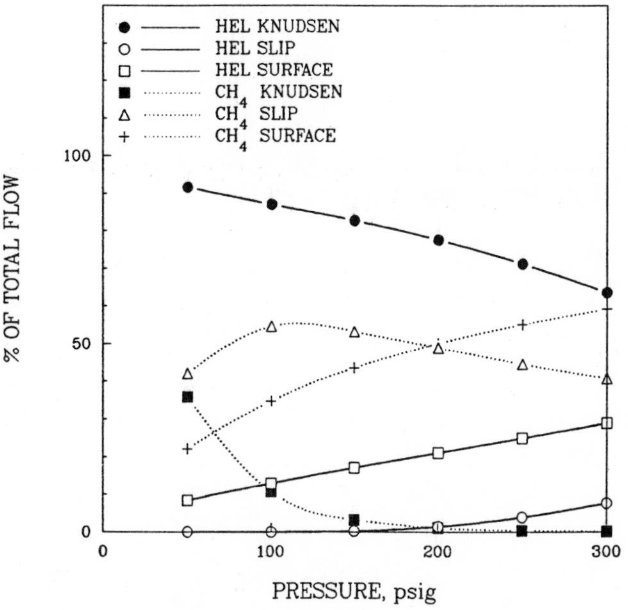

Figure 6. Contribution of flow components to the total flow of helium and CH_4 for the membrane mounted in cell (4-8).

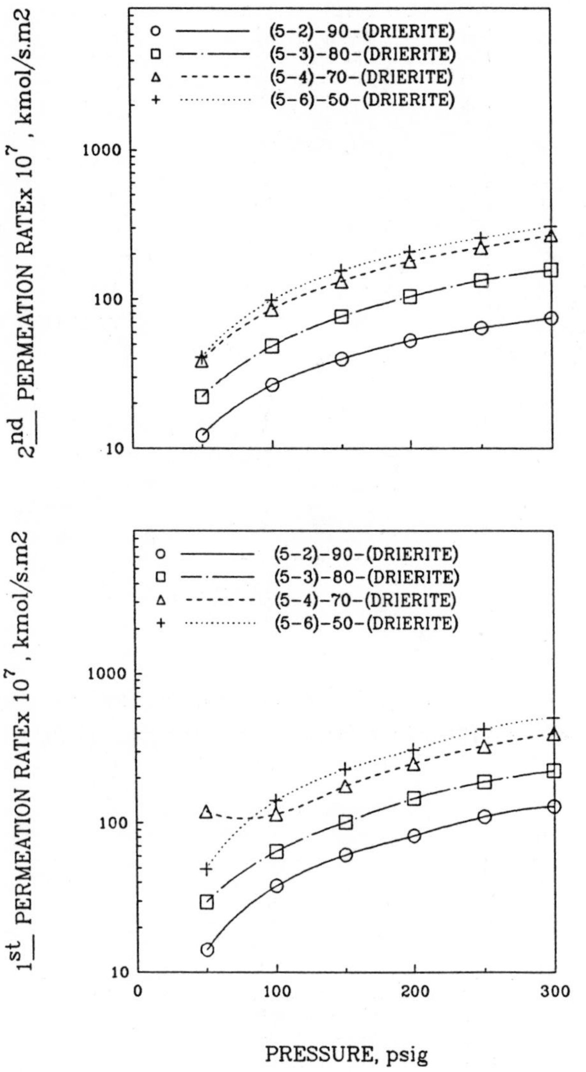

Figure 7. Effect of shrinkage temperature on permeation rates of CO_2 for membranes in group (5).

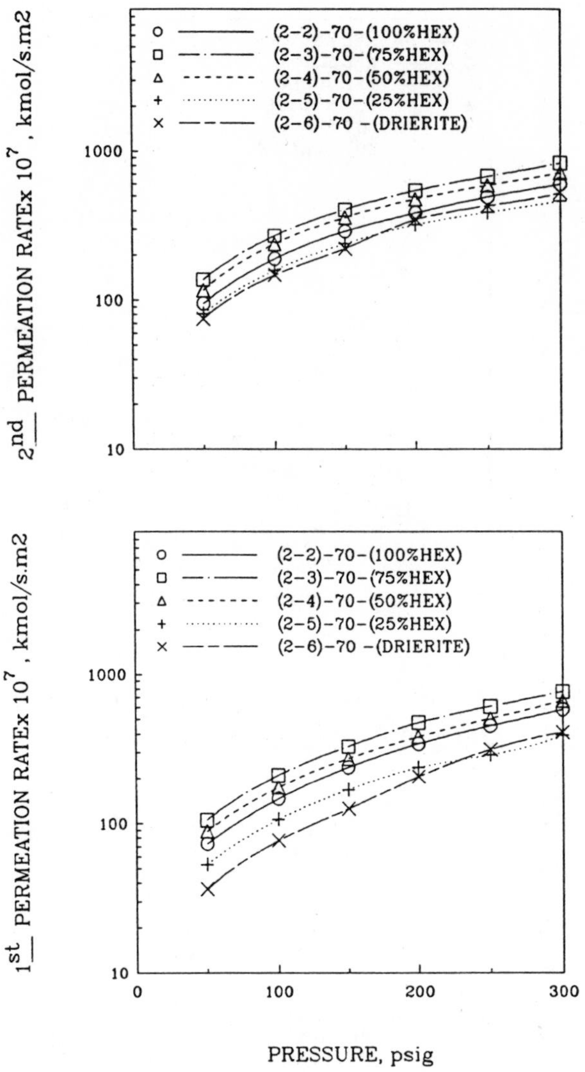

Figure 8. Effect of vapor pressure on permeation rates of helium for membranes in group (2).

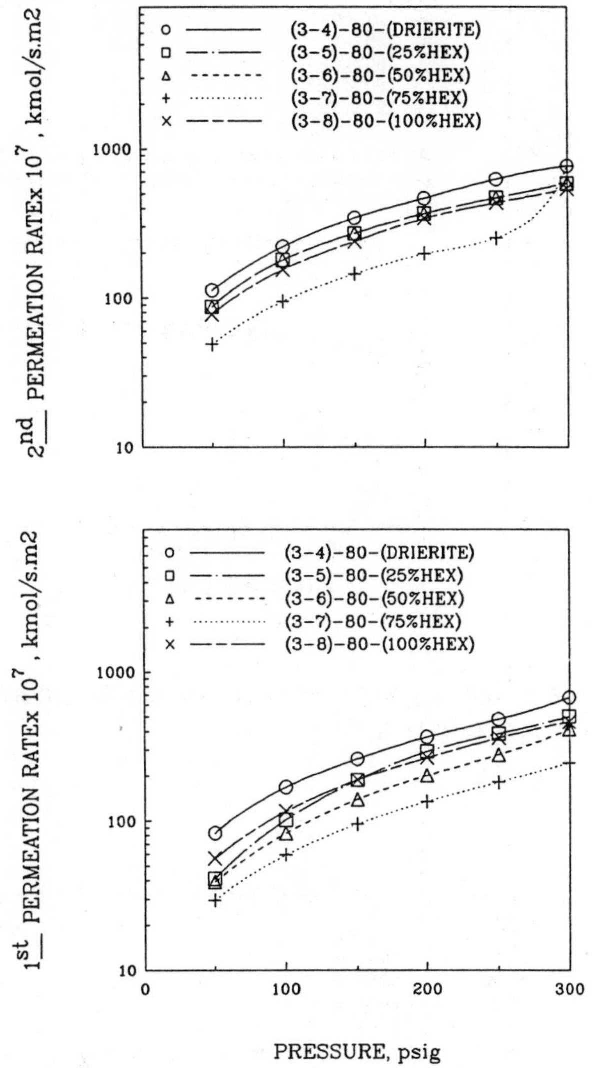

Figure 9. Effect of vapor pressure on permeation rates of helium for membranes in group (3).

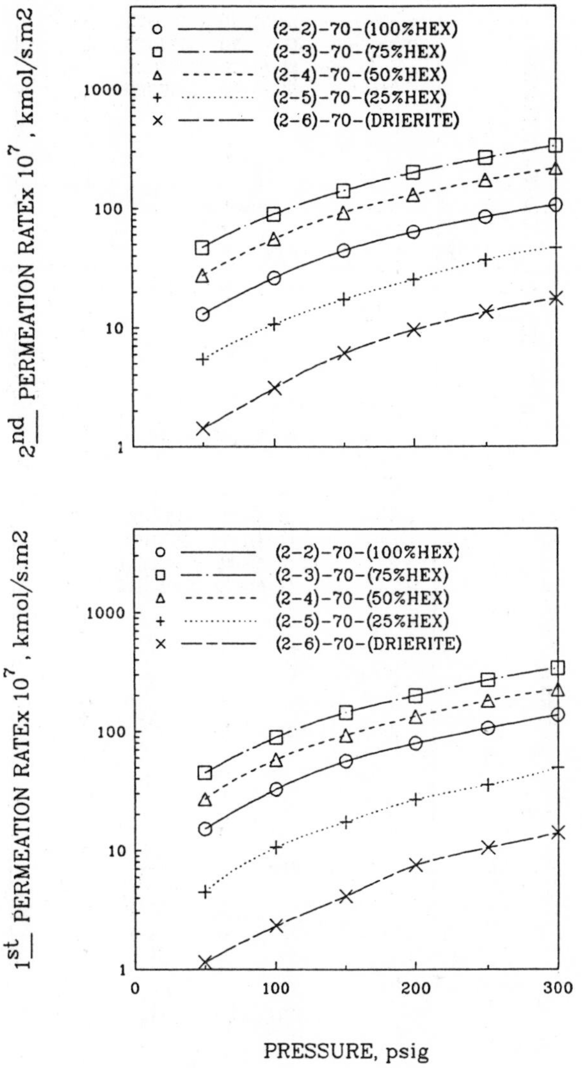

Figure 10. Effect of vapor pressure on permeation rates of CH_4 for membranes in group (2).

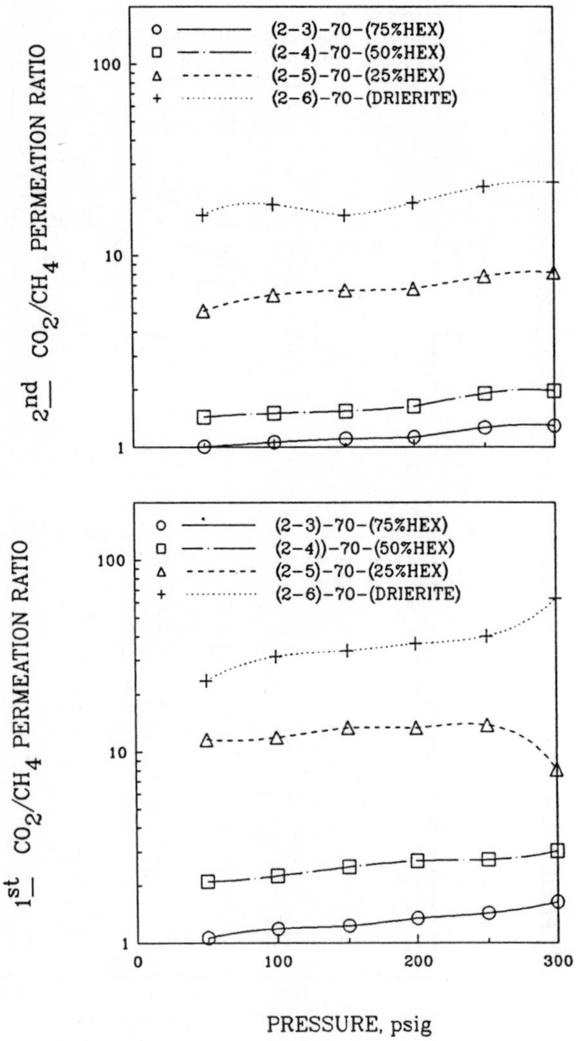

Figure 11. Effect of drying conditions on CO_2/CH_4 permeation ratio for membranes in group (2).

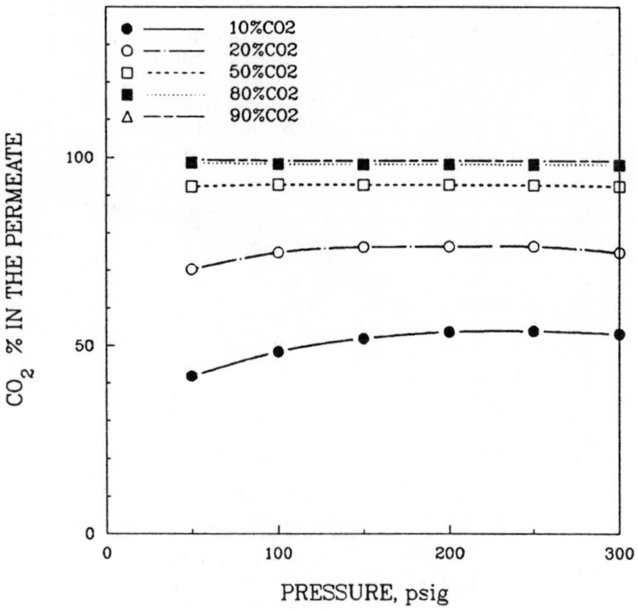

Figure 12. Effect of inlet pressure on CO_2 % in the permeate for membranes in group (5).

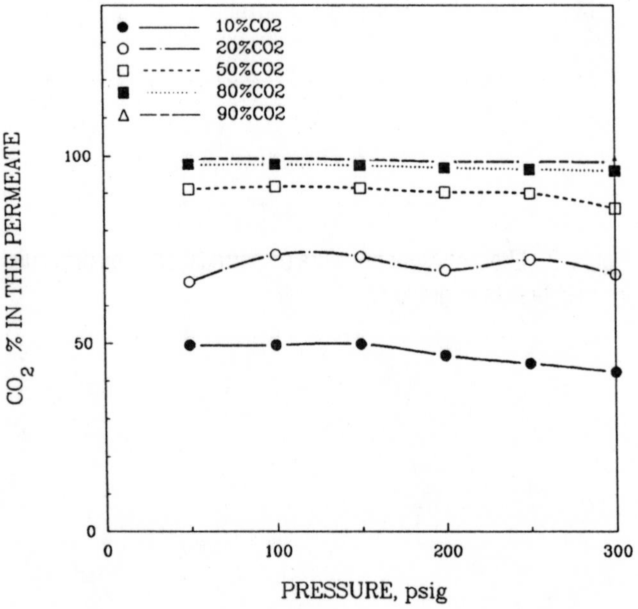

Figure 13. Effect of inlet pressure on CO_2 % in the permeate for membranes in group (1).

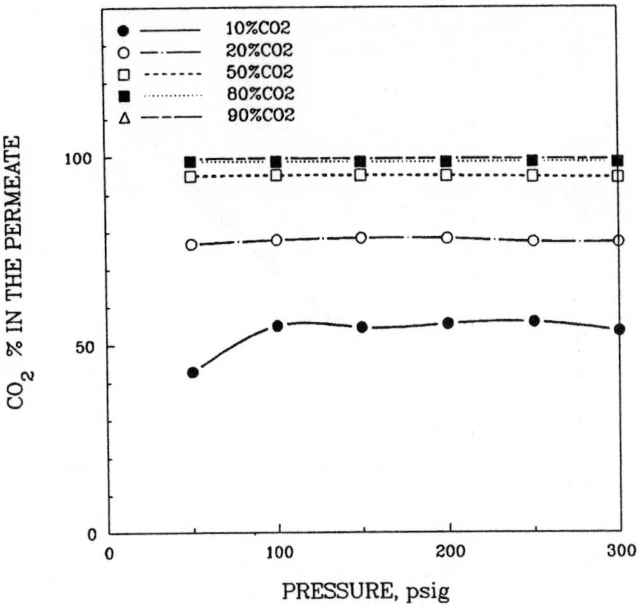

Figure 14. Effect of inlet pressure on CO_2 % in the permeate for membranes in group (6).

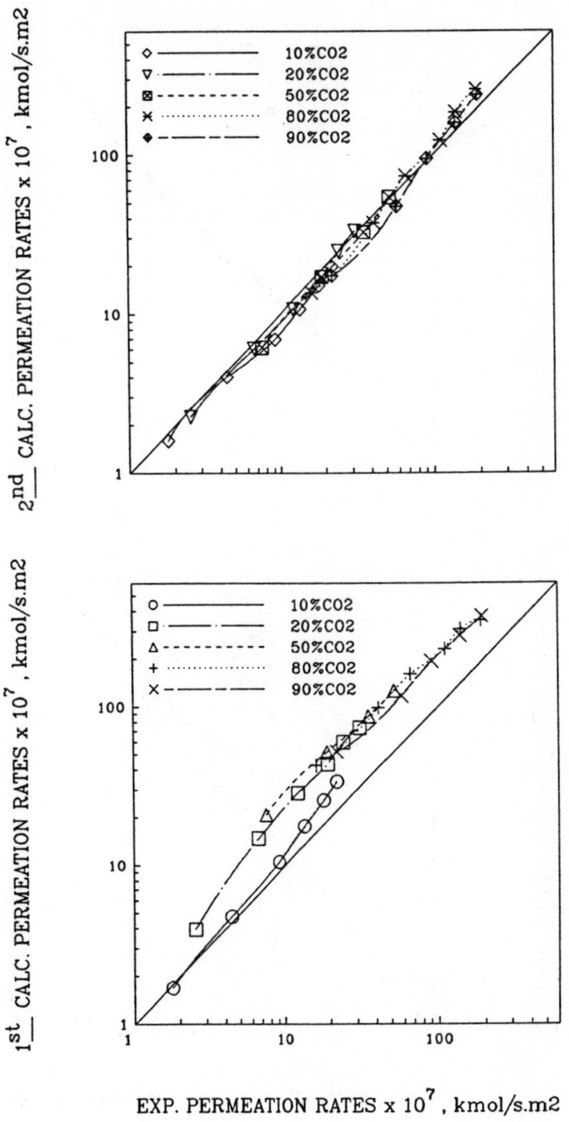

Figure 15. Effect of characterization time on the permeation rate prediction for membrane mounted in cell (2-6).

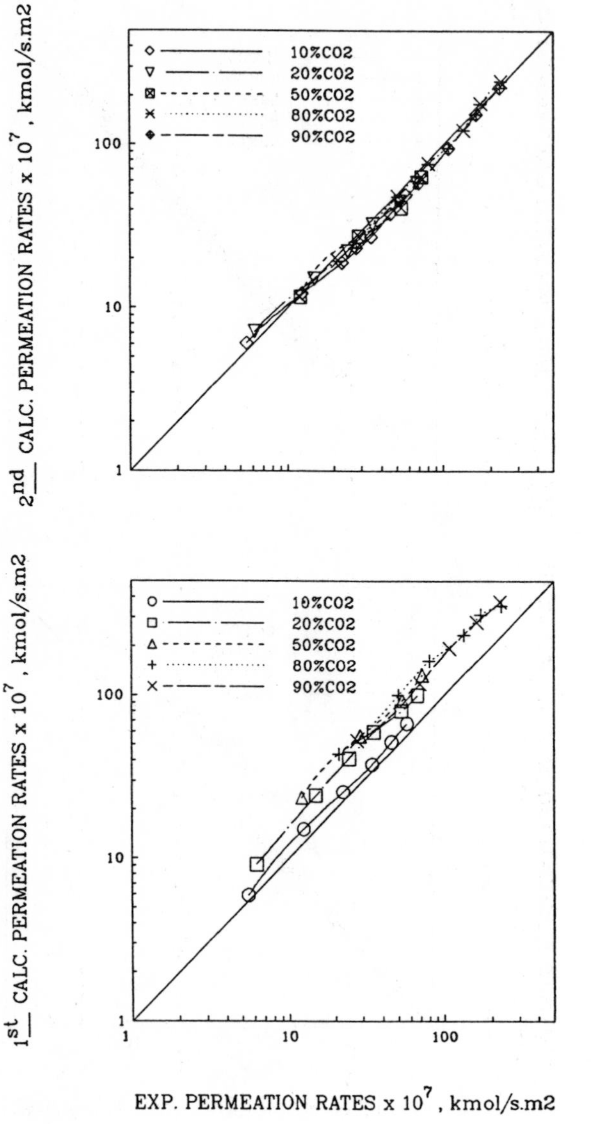

Figure 16. Effect of characterization time on the permeation rate prediction for membrane mounted in cell (2-5).

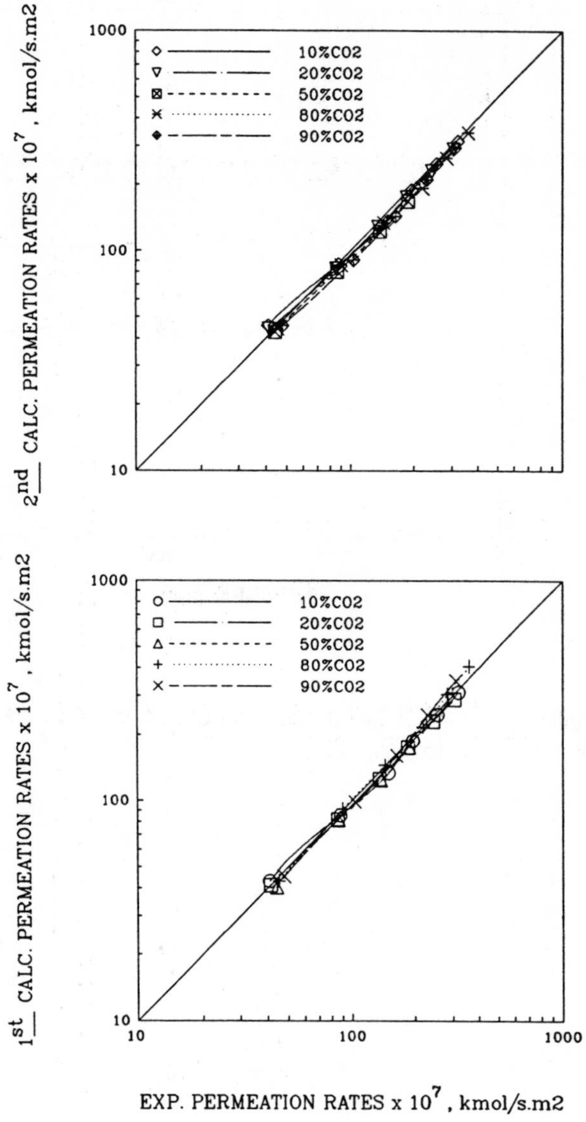

Figure 17. Effect of characterization time on the permeation rate prediction for membrane mounted in cell (2-3).

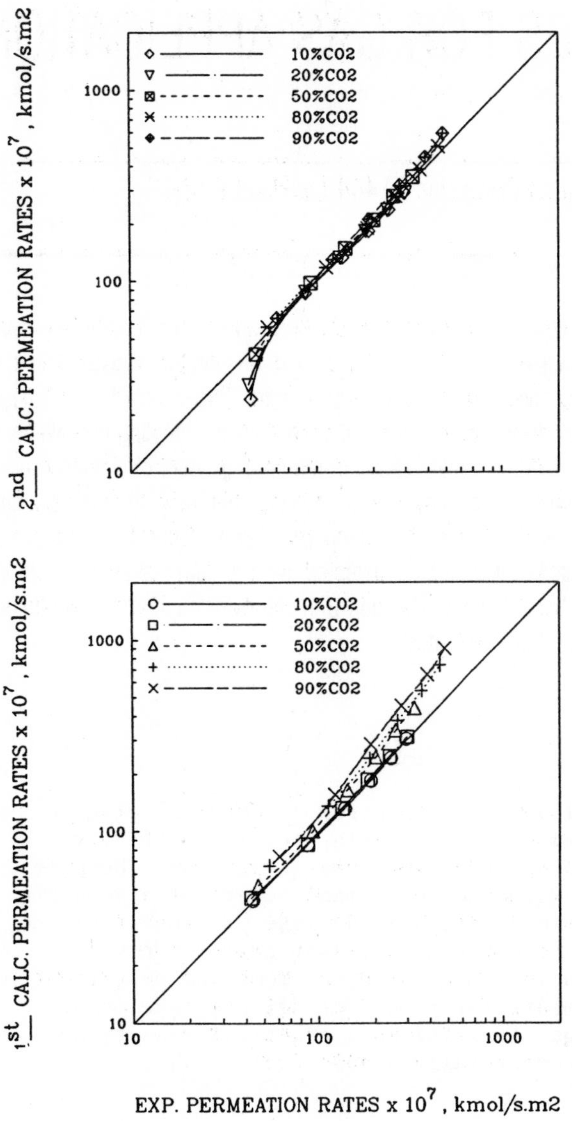

Figure 18. Effect of characterization time on the permeation rate prediction for membrane mounted in cell (1-3).

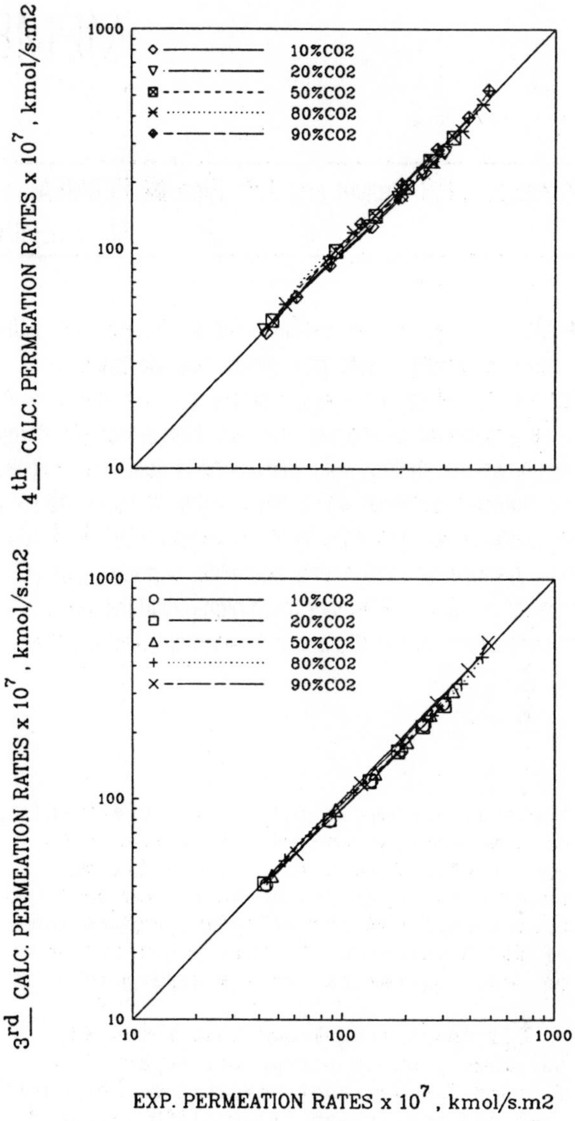

Figure 19. Effect of characterization time on the permeation rate prediction for membrane mounted in cell (1-3).

POST-TREATMENT OF ASYMMETRIC MEMBRANES FOR GAS APPLICATION

M.K. Murphy, E.R. Beaver and A.W. Rice ■ PERMEA, Inc., A Monsanto Company, 11444 Lackland Road, St. Louis, MO 63146

In the mid-1970's, the development of the Prism® multicomponent composite membrane by Monsanto/Permea ushered in the era of commercially viable gas separations via membranes. Key to this development was a patented membrane post-treatment by which minute defects or pores, typically found in the ultra-thin separating layer, were repaired or sealed. This enabled simultaneous achievement of high permeation rates and high selectivity. Permea has continued to advance the state of the art by coupling improvements in membrane morphology with innovations in post-treatment technology. In 1986, a breakthrough in morphology control yielded Prism® Alpha Nitrogen systems which have higher transport rates while maintaining selectivity. More recently, the new Prism® Cactus® Dryer systems for gas dehydration have exploited novel post-treatments whereby the membrane porosity is tailored to achieve an optimal balance of design and operating characteristics. Other post-treatment technology patented by Permea makes possible the attainment of separation factors much higher than the intrinsic values for the base polymer. Coupled with improved membrane morphology, such permeation modifications can be accomplished at economically practical transport rates.

Physical, chemical and thermal treatments or combinations thereof offer effective and practical ways to flexibly tailor transport properties of preformed membranes employed for selective separations of fluid mixtures. In this paper we focus on post-treatments for gas separations.

Although long known that selected polymers provide attractive separation factors for various separations, not until Loeb and Sourirajan (1) in 1960 invented integrally skinned asymmetric membranes, were permeation rates high enough for practical utility possible. That new membrane structure gave an ultra-thin separating layer, for high rates, supported by a mechanically strong but nonselective porous substructure. Success of cellulosic asymmetric membranes in reverse osmosis desalination attested to the revolutionary nature of their invention.

And yet, gas separations based upon asymmetric membranes did not enjoy so immediate a reduction to commercial practicality. High rates were similarly necessary, but another problem remained before practical gas separation membranes were to become a reality. As separating layers were made ever thinner to increase rates toward practically useful levels, detrimental effects of imperfections in the separating layer (minute pores or defects) plagued attempts to obtain useful selectivity properties for separating gases. Despite extreme care in manufacture of asymmetric membranes, such defects are inevitable to a greater or a lesser degree. Only the development of novel membrane post-treatments for repair of defects made membrane gas separations a practical reality on commercially viable scale.

REPAIR OF DEFECTS IN ASYMMETRIC GAS SEPARATION MEMBRANES

In the mid-1970's at Monsanto, Henis and Tripodi (2) developed defect repair post-treatment technology which made possible the first economically viable large scale membrane gas separations. These multicomponent composite membranes are asymmetric polymeric membranes, with minute defects in the active separating layer repaired by occlusive plugging of defects with a material of higher permeability but lower selectivity relative to active separating layer polymer.

Operative effect of defect plugging is to increase resistance to transport through pores or defects, forcing gases to transport through the selective polymer layer comprising the majority of active membrane surface area. Occlusion or plugging of defects eliminates nonselective transport pathways. Selective transport is controlled by permeability and selectivity characteristics of membrane polymer. This provides a simple, yet efficient and generally applicable solution to the problem of obtaining high rate and high selectivity and has been successfully applied to a variety of commercially important applications of membrane gas separations.

A wide range of occlusive plugging materials have proven effective and functionally equivalent with respect to improving selectivity, from low molecular weight oils to high molecular weight polymers. Practical considerations, such as aggressiveness of field conditions and ease of routine manufacturing often dictate choice of materials and method of treatment. In some situations, surface active components provide improved adhesion depending upon the nature of the membrane polymer.

Basic relationships governing such post-treatment are enunciated in a mathematical formalism, called the Resistance Model (2, 3). Analogy is made to a combination of series and parallel resistances in electrical circuits. The model has been refined by Altena and Henis (4) to include non-negligible resistance in the porous support substructure.

Figure 1 depicts occlusive plugging of membrane defects and provides essentials of the Resistance Model formalism. Parts b and c of the figure depict two extremes of occlusive material deposited onto the membrane. Permeation rate, or flux, for component i through a polymeric membrane can be described by: $Q_i = P_i A \Delta c_i / \ell$ where Q_i is permeation rate of component i, P_i is intrinsic permeability of membrane polymer to component i, A is the surface area of membrane available for permeation, ℓ is thickness of membrane through which component permeates, and Δc_i is concentration difference of component i across the membrane (in the case of gas permeation, the component partial pressure difference).

Using the electrical analogy, the above relationship describing permeation through a polymer membrane is mathematically equivalent to Ohm's law which describes current flow through a resistor: $I = E/R$. Conceptually current, I, can be equated with permeation rate, Q. Driving force for current flow, (i.e. difference in electric potential E), is analogous to concentration gradient Δc_i. Then, resistance to permeate flow, R_i, is equivalent to electrical resistance R: $R_i = \ell / P_i A$, so combining $Q_i = \Delta c_i / R_i$.

Figure 1b reflects a situation after deposition of high molecular weight polymeric occlusive material from relatively high concentration solution in volatile solvent carrier. Occlusive polymer plugs defects and persists as a moderately thick coating overlaying much of the membrane surface. Since it does not contribute to selective separation of gases, such a layer may be of little practical consequence, as long as it is highly permeable and does not contribute significant series resistance R1 to total gas transport resistance Rt.

Figure 1c reflects the other extreme, where only plugging of pores or defects occurs with no deposition of occlusive material over remaining defect-free regions of membrane surface. Total resistance is determined by relative parallel resistances of plugged defects R3 and defect-free region of separating layer R2.

In both instances, 1b and 1c, occlusive plugging material serves to increase R3 relative to R2, thus requiring membrane polymer transport properties to dominate separation characteristics of the post-treated membrane. Any post-treatment material which occludes or plugs pores or defects can be effectively used, provided it does not control overall membrane resistance and transport characteristics, i.e. R1 does not dominate Rt.

Defect repair post-treatments are equally effective for a variety of membrane systems, wherever minute defects or pores otherwise short circuit high selectivity intrinsic to polymeric material of the thin separating layer. This technology is applicable to asymmetric membranes in flat and hollow fiber form and to ultra-thin separating layers in composite membranes for gas separations (5). A pore or other minute defect, its detrimental effect on selectivity, and repair by occlusive plugging are functionally identical in all membranes for gas separation applications. Common feature is elimination of low resistance, nonselective pathways for gas to traverse the membrane, thus shunting gas transport through the intended selective polymeric active layer of the membrane.

In 1986, development of methodologies for improvements in membrane morphology yielded a new generation of high productivity asymmetric membranes. Fundamental structural characteristics and methods of preparation have been presented previously and will be further discussed in detail elsewhere (6-8). The essential feature of the new membranes lies in substantial increases in gas transport rates, while selectivity is maintained at intrinsic levels of the polymer.

Commercial applications of the new membranes have focussed on separations leveraged by much higher rates, including applications for gases of relatively low value or slower intrinsic permeability and where separator size or weight is important. Higher rate membranes employed in Prism® Alpha gas separation systems afford lowered system capital and operating cost per unit of gas product than prior slower membranes. Successful commercial applications include generation of nitrogen from air for a variety of inerting, blanketing and controlled atmosphere applications (9-12).

Most pertinent to present discussion, post-treatment of high rate membranes, though entailing some process differences, is based essentially upon the defect repair, occlusive plugging post-treatment technology discussed above. High rates of these new membranes present opportunities beyond above mentioned applications. Novel post-treatments of different types have been employed to tailor these new membranes for other gas separation applications, which will be the focus of the remainder of this discussion.

Many researchers in the field of gas separation membranes have and continue to focus on modifications to polymer molecular structure and search for polymers having unique intrinsic transport properties. This approach to the ideal makes good sense from an academic research viewpoint, to utilize the optimum polymer for each separation. From a pragmatic industrial perspective of membrane and systems manufacturer, capability to effectively and controllably tailor a given polymer membrane, alter performance characteristics, and optimize transport properties specifically designed to match in each instance a number of important commercial gas separation applications offers unique flexibility. Appropriate post-treatment technologies can provide that unique flexibility and we continue to pursue that practical ideal.

POROSITY CONTROL IN ASYMMETRIC GAS DEHYDRATION MEMBRANES

Recently commercialized new membrane gas dehydration technology, aimed at drying compressed air for use in various pneumatic applications, called Prism® Cactus® Dryer systems, are based on hollow fiber membranes post-treated to alter membrane porosity (13). Control of pore size and distribution results from specific combinations of chemical and thermal post-treatments to preformed precursor membranes.

Such post-treatments yield an optimal balance of system design and operating characteristics. High water vapor transport and controlled air transport rates through porous hollow fiber membranes provide both effective drying of product nonpermeate air and effective sweep of permeated water vapor from downstream or permeate side of the fibers. Controlled sweep maintains maximum transmembrane partial pressure differential and thus driving force for water transport, with a minimum of system complexity and feed air expenditure.

Such systems produce dried product air of 0°C pressure dewpoint, from feed compressed air of 40°C dewpoint, operating at feed pressures up to 2070 KPa (300 psig). Gas flow depends on membrane area, with commercial systems ranging up to 1.1 Nm^3/min (40 SCFM) dried air product, from module sizes ranging up to 7.5 cm (5 in.) diameter by 68 cm (25 in.) long. Pressure drop across the module is less than 13.8 KPa (2 psig), thus dry product gas is at essentially the pressure of feed.

Optimal balance of transport properties for water vapor and air through controlled porosity membranes is depicted schematically in Figure 2, where membrane permeability of water vapor, P/ℓ H_2O, is plotted versus permeability for air, P/ℓ Air.

Central portion of this graph is a region of optimal property combinations for compressed air dehydration applications, suitable combinations of water vapor permeability relative to air permeability, which yield optimized membrane system performance. Four regions of the graph outside the bounds of this central region characterize combinations of transport properties much less suitable for membrane air dehydration.

Region A, high P/ℓ H_2O and low P/ℓ Air: a membrane which provides inadequate sweep to remove permeated moisture from downstream side of membrane, thus failing to maximize driving force for water transport. Alternative recycle of dry product to provide required sweep of permeate side of membrane leads both to increased hardware complexity and significant sacrifice of product flow.

Region B, high P/ℓ H_2O and high P/ℓ Air appears attractive at first. But detailed analyses of system design variables for this region suggest significant practical problems, such as excessive system pressure drop. Fabrication of very large diameter modules of very short length for operation in this region poses severe difficulty in engineering and manufacturing. Such a system package limits effective removal of permeated moisture.

Region C, low P/ℓ H_2O and high P/ℓ air: poor drying performance and excessive loss of feed air to permeate. Region D, low P/ℓ H_2O and low P/ℓ Air: systems requiring very large membrane areas and excessive capital.

The central region is bounded by the following combinations of H_2O and air permeabilities: P/ℓ H_2O about 300 to 1500 standard units and P/ℓ Air about 10 to 100, provided that the selectivity H_2O/Air or the ratio of respective permeabilities is in the range of about 10 to 50. Permeability units are 10^{-6} cm^3(STP) / cm^2 x sec x cm Hg. That these membranes are porous is seen from limited selectivity for O_2/N_2 separation, typically in the range of only 1.05 to 2.0, for membranes constructed of polymers having intrinsic O_2/N_2 selectivities of greater than 4.

Post-treatments suitable for achieving such optimal balance of properties for air dehydration include gentle chemical annealing with organic liquids, which have swelling effects on but are not strong solvents for the polymer of the membrane. A useful example, for treatment of polysulfone membranes, is methyl alcohol.

Thermal post-treatments are also suitable for porosity adjustment. Examples of chemical and thermal post-treatments for controlled adjustment of membrane porosity may be found in reference (13). Polymers as diverse as polysulfone, polyphenylene oxide, acrylonitrile-co-styrene and aromatic polyamides, all in asymmetric hollow fiber membrane form are included.

Figure 3 shows hardware arrangements for effective application of controlled porosity membrane dryers in two common situations; first, use with dedicated air compressor, second, point of use remote from the compressor. In large plant-wide compressed air utilities, dryers perform best installed at point of use. Typical module performance of commercial air dryer systems is shown in Table 1, as a function of module size for operation at 1035 KPa (150 psig) and at 2070 KPa (300 psig) feed pressures.

PERMEATION MODIFICATION FOR HIGHER THAN INTRINSIC SELECTIVITY

A prime target among researchers seeking new, unique polymers for use in gas separation membranes is high productivity: a combination of high permeability and high selectivity. In many glassy polymers, intrinsic properties fall disappointingly short of this target. Among most polymers, higher permeability is accompanied by lower selectivity. Similarly, highly selective polymers frequently have low permeabilities.

Recently some progress has been made in design of polymers which combine both high permeability and selectivity. Notable are novel polyimides, with unique combinations of macromolecular chain rigidity, structural bulkiness and chain segmental mobility (14, 15). If these materials can be economically translated into suitable membranes, either asymmetric or composite structures, one would expect them to find utility in gas applications.

As discussed earlier, advantages in practical utility can also be had by appropriate post-treatments of preformed asymmetric membranes. Incorporation of permeation modifier compounds into the thin separating layer of asymmetric membranes can provide selectivities increased to levels significantly higher than intrinsic for a variety of membrane polymers (16-18). Increases in selectivity occur at expense of permeability. However, control of treatment conditions, such as choice of modifier and application mode, time of exposure, post-incorporation thermal treatment, and most critically choice of high permeability precursor membranes yields systems with economically practical transport rates.

Permeation modifiers or antiplasticizers are generally polar, low to moderate molecular weight materials. When incorporated into glassy polymers at low concentration, such compounds increase stiffness, usually decrease glass transition temperature and in some cases increase selectivity (19, 20).

Maeda and Paul reported on effects of some modifiers with certain glassy polymers, including polysulfone and polyphenylene oxide (19, 20), effects on gas sorption, transport and polymer free volume, and provided fundamental insights into such effects.

We have focused extensive efforts on use of this technology to increase selectivity of polymers in asymmetric hollow fiber membrane form (16-18). Dramatic selectivity increases occur for gas pairs differing significantly in molecular size. Similar size permeant pairs display smaller, yet useful increases in selectivity. Paul showed such effects to be related to the additive's principal influence on permeant diffusivity, rather than solubility, in the glassy polymer.

Post-treatments which incorporate modifier into only the thin separating layer of the membrane avoid detrimental reductions of the polymer's glass transition temperature. Significant selectivity increases are observed at modifier concentrations of 10-30% relative to membrane polymer weight. However, limiting incorporation to only the thin separating layer of asymmetric membranes results in large selectivity increases with very low levels of modifier relative to overall membrane weight, significantly less than 0.1%. Calorimetric Tg measurements of modified asymmetric membranes detect no reduction in overall fiber Tg. Thus, key mechanical properties, such as pressure capabilities which rely on integrity of the membrane's porous substructure, remain unaffected.

A straightforward post-treatment process for asymmetric hollow fiber membranes employs application of modifier from dilute solution in volatile solvent. Concentration and time of exposure can be varied to increase amount of modifier incorporated and thus the degree of selectivity increase.

Data in Table 2 for helium/nitrogen separation demonstrate the effect of one compound, 2-ethyl,4-methyl imidazole, on hollow fiber membranes of polysulfone and polyethersulfone. Additional examples may be found in references (16-18). Precursor membranes employed are new high rate membranes, prepared using Prism® Alpha technology alluded to above (6-8). Permeation modified membrane properties are compared in the table with untreated versions of slower first generation Prism® membranes, previously proven to have wide practical commercial utility.

Data show dependence of membrane permeability and selectivity on extent of treatment. Higher concentration of modifier application solution yields higher He/N_2 selectivity and lower P/ℓ He. Note that selectivity is increased by 50% to 150% above that of respective untreated membranes and that when the precursor membrane is very fast initially, trade-off of reduced permeability leaves modified membranes with higher rates than unmodified first generation membranes.

Examples presented in Table 2 employed application of modifier from dilute solution in volatile solvent, followed by defect repair post-treatment. Modifiers can be incorporated directly into the defect repair process, either by simple mixing with defect repair material or by covalent attachment to polymeric defect repair materials (18). In such instances, increased stability and improved adhesion are possible.

CONCLUSIONS

Post-treatments of preformed asymmetric polymer membranes offer substantial practical benefits for tailoring gas transport properties of membranes and can be effectively employed to render a given precursor membrane suitable for use in a wide range of commercially important gas separation and purification applications.

The diversity of transport properties attainable range from control of membrane porosity for dehydration, to defect repair by occlusive plugging of pores for achievement of intrinsic selectivity, to elevation of membrane selectivity to levels significantly greater than intrinsic levels characteristic of the precursor polymer. Post-treatment technology is effective for a wide range of polymers in membrane form.

Membranes of greatest practical utility combine appropriate post-treatment technology with the proper precursor to obtain optimum combinations of rate and selectivity in the final membrane. This approach offers important practical advantages to development of a unique polymer for each gas separation application.

LITERATURE CITED

1. Loeb, S. and S. Sourirajan, Report No. 6060. Dept. of Engineering, University of California, Los Angeles, CA (1960).

2. Henis, J. M. S. and M. K. Tripodi, U.S. Patent 4,230,463.

3. Henis, J. M. S. and M. K. Tripodi, J. Membrane Sci., 8, 233 (1981); and Sep. Sci. and Tech., 15 (4) 1059 (1980).

4. Altena, F., A. A. Brooks, M. K. Tripodi and J. M. S. Henis, unpublished results.

5. Kraus, M. A. and C. N. Tran, U.S. Patent 4,806,189.

6. Kesting, R. E., A. K. Fritzsche, M. K. Murphy, A. C. Handermann, C. A. Cruse and R. F. Malon, U.S. Patents pending.

7. Fritzsche, A. K., M. K. Murphy, C. A. Cruse, R. F. Malon and R. E. Kesting, "Asymmetric Hollow Fiber Membranes with Graded Density Skins", presented at AIChE Meeting, Washington, D.C. (1988).

8. R. E. Kesting, A. K. Fritzsche, M. K. Murphy, C. A. Cruse, A. C. Handermann, R. F. Malon and M. D. Moore, to be published.

9. Bhat. P. V. and E. R. Beaver, AIChE Symp. Series, 84 (261) 124 (1988).

10. Fritzsche, A. K., Polymer News, 13 (9) 266 (1988).

11. Beaver, E. R., V. Bhat and D. S. Sarcia, AIChE Symp. Series, 84 (261) 113 (1988).

12. Murphy, M. K., A. W. Rice and J. J. Freeman, Spectroscopy, 3 (10) 32 (1988) and U. S. Patent 4,793,830.

13. A. W. Rice and M. K. Murphy, U.S. Patent 4,783,201.

14. (a) Kim, T. H., W. J. Koros, G. R. Husk and K. C. O'Brien, "Relationship Between Gas Separation Properties and Chemical Structures in a Series of Aromatic Polyimides", presented at AIChE Meeting, Houston, Texas (1987); (b) Kim, T. H., W. J. Koros and G. R. Husk, "Advanced Gas Separation Membrane Materials: Rigid Aromatic Polyimides", presented at Fifth Symp. on Separation Sci. and Tech. for Energy Applications, Knoxville, Tennessee (1987).

15. Hayes, R. A., U.S. Patents 4,705,540, 4,717,393 and 4,717,394.

16. Brooks, A. A., J. S. Fried, J. M. S. Henis, A. Zampini and D. Raucher, U.S. Patent 4,575,385.

17. Murphy, M. K., U.S. Patent 4,728,346.

18. Zampini, A., U.S. Patent 4,484,935.

19. Maeda, Y. and D. R. Paul, J. Polym. Sci., (B) Polym. Phys., 25, 957 (1987); ibid., 25, 981 (1987); and ibid., 25, 1005 (1987).

20. Maeda, Y. and D. R. Paul, J. Membrane Sci., 30, 1 (1987).

Table 1 Typical Performance for Drying Air using Prism® Cactus® Membrane Systems

Feed Air Pressure (KPa)	Module Dimensions Length (cm)	Diameter (cm)	Product Capacity (Nm^3/min)	Module Weight (kg)
1035.	35.6	6.4	0.05	0.9
	63.5	6.4	0.11	1.4
	38.1	8.9	0.24	2.3
	64.7	8.9	0.51	3.2
2070.	35.6	6.4	0.11	2.3
	62.2	6.4	0.21	2.7
	36.8	12.7	0.51	6.4
	63.5	12.7	1.07	7.7

Product flow values assume product is dried to 0°C pressure dewpoint, from feed air of 40°C dewpoint at indicated feed pressure. Capacity depends on feed pressure and feed/product dewpoint and can be more than twice typical values.

Table 2 Permeation Modified Asymmetric Membranes for Gas Separation Applications

Membrane Polymer	Modifier Concentration (wt%)		Permeability P/ℓ Helium (10^{-6})	Selectivity (He / N_2)
Prism® Alpha-type:				
Polysulfone	0	(a)	170.	91.
	0.5	(b)	103.	163.
	2.0	(b)	73.	247.
Polyethersulfone	0	(a)	84.	182.
	0.5	(b)	46.	308.
	2.0	(b)	30.	366.
Prism-type:				
Polysulfone	0	(a)	75.	90.
Polyethersulfone	0	(a)	32.	131.

Notes: (a) Unmodified/defect repair post-treated with 1% Sylgard in isopentane
(b) Modifier: 2-ethyl, 4-methyl imidazole applied from solution in (98/2 wt/wt) cyclohexane/methyl alcohol

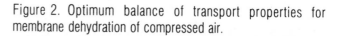

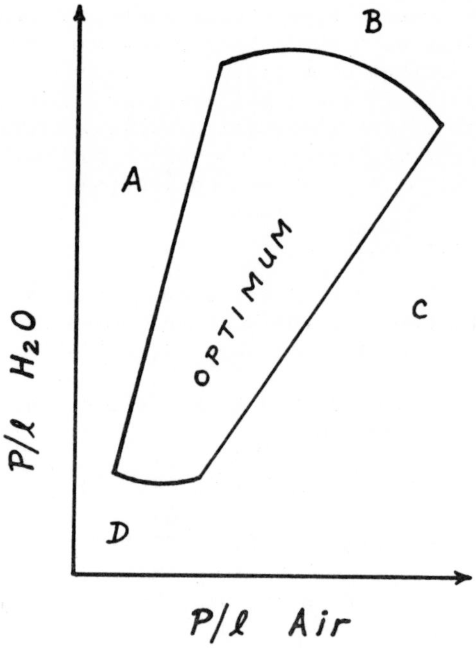

Figure 2. Optimum balance of transport properties for membrane dehydration of compressed air.

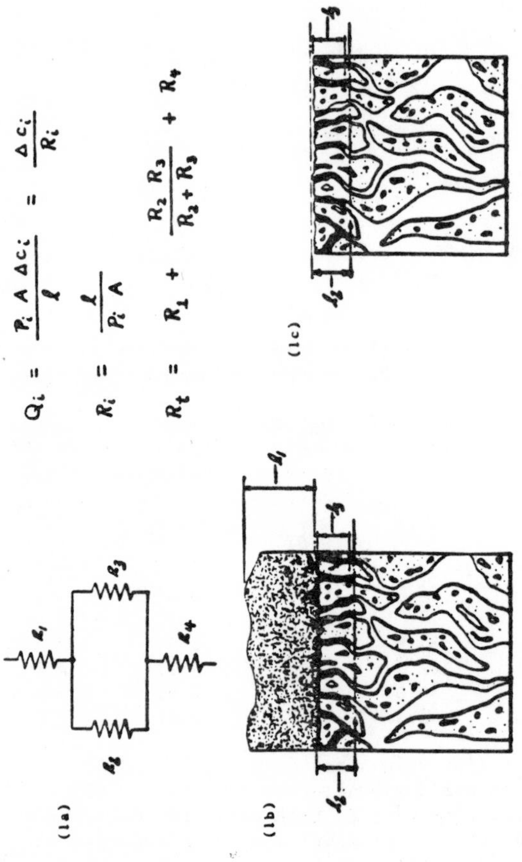

Figure 1. Resistance model for occlusive defect repair post treatment.

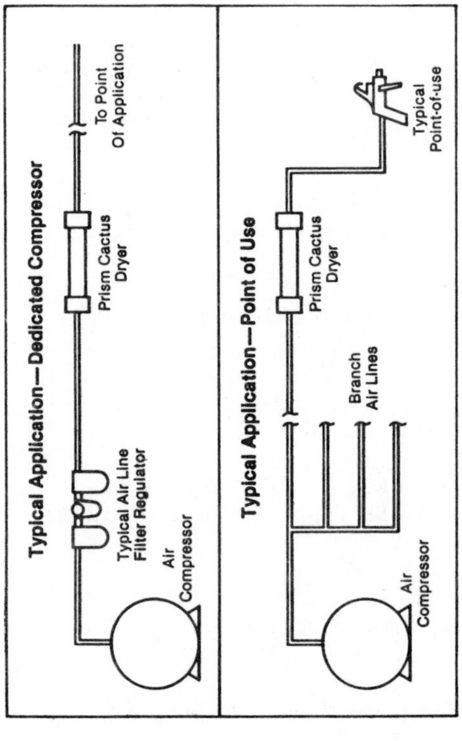

Figure 3. Typical application modes for membrane dehydration of compressed air.

ANALYSIS AND DESIGN OF TWO-MEMBRANE COLUMN FOR GAS SEPARATIONS

Zhiquan Yan and Yuen-Koh Kao ■ Department of Chemical and Nuclear Engineering, University of Cincinnati, Cincinnati, OH 45221

Given a specific feed gas mixture and membrane materials, the key design variable for a Two-Membrane Column (TMC) is the area ratio of the two membranes in a column. The advantage of the TMC design is fully utilized only when membrane area allocations for each stage are properly adjusted. An optimally designed TMC showed significant improvements in separation efficiency for the separations of helium-hydrocarbon mixtures both in product purity and in Rony's extent of separation index. Effects of feed composition and membrane selectivity on column performance were presented. The relative area of the membrane in the second stage must be increased with increasing feed composition to maintain the optimum performance of the TMC. The separation of a binary gas mixture by a TMC could be enhanced by using membranes with high reversed selectivities.

Membrane gas separation as a convenient, energy saving process is now competing with cryogenics and a variety of adsorption and absorption processes (1). For some applications, a simple single-stage membrane permeator is adequate. However, as the size of the process grows or as the value of the separated components increases, multistage design arrangements can be beneficially employed. Engineering design of membrane systems employing complex configurations is currently under extensive investigation to improve the membrane separation efficiency. Among many design alternatives suggested and studied (2-10), Two-Membrane Column (TMC), which employs two membranes with opposite selectivities in an arrangement as shown in Figure 1, has been advanced as an energy-saving, high purity separation process for binary gas separation (2). Comparative studies have shown that TMC can attain higher separation efficiency over other two-membrane system designs (3).

In a single membrane permeator, the permeate composition has an upper limit determined by the selectivity of the membrane-gas pair. While the permeate stream cannot provide high purity of more permeable species, the reject stream can produce an end product of highly-purified less permeable species. The TMC design utilized this feature by collecting both reject streams of its two stages to obtain highly concentrated products at both ends of a column.

Most previous work on TMC was directed toward the experimental verification and the optimization of the process was not addressed (2,4). Results from the experimental studies using TMC of a predetermined dimension may not fully reveal the optimum performance of the TMC. As will be shown in this study, TMC may be most advantageously used only when it is properly designed. The optimum amounts of membrane areas to be used in each stage of the TMC for a given separation problem need to be determined to achieve its best performance. Since economics of a membrane process can be significantly affected by the design of the process (1), detailed analysis and optimal design of the TMC is essential for its future economic studies and for its potential applications.

The present study focuses on the optimal utilization of the TMC for binary gas separations. A comprehensive analysis of how process optimization can affect process performance is presented. Results of parametric study are also presented to illustrate the effects of feed composition and membrane selectivities on the TMC performance.

SIMULATION METHOD

The following simulation utilized the countercurrent "plug-flow" model equations based on material balances. Detailed development of the differential equations describing the permeator modules can be found in other references (2,11,12,13). In general, the model for a TMC consists of four coupled, nonlinear ordinary differential equations with boundary

Correspondence concerning this paper should be addressed to Y-K Kao.

conditions specified on the opposite ends of the permeators. The orthogonal collocation method (14) was used to solve the problem numerically. This method reduced the differential equations and the boundary conditions to a system of algebraic equations which were solved simultaneously using the *IMSL* algebraic-solving routine *ZSCNT*. Details on the numerical scheme can be found elsewhere (15).

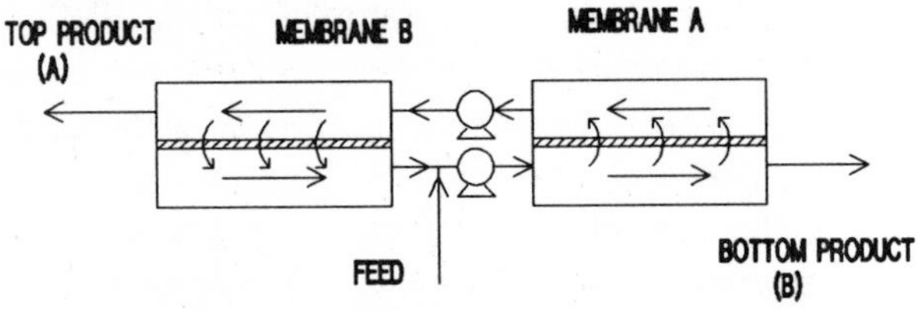

Figure 1. Two-Membrane Column (TMC)

In the analysis of a membrane permeator, the following design and process variables are important: the membrane properties (area, thickness, permeability and selectivity), the separation requirements (feed rate and feed composition, product purity and recovery, and separation efficiency), and the operating conditions (pressure and temperature). Performance criteria are chosen according to the specific separation requirements. In many situations, high purity product is required. In other situations, the best compromise between product purity and recovery is needed to maximize the overall extent of separation. The key design variables for all cases are the specific areas of the respective membranes to be used in each stage of a TMC. There exists a unique minimum total membrane area and an optimum membrane area ratio for a specified separation task.

The example chosen for the present study was the common gas separation problem in helium recovery application, i.e., the separation of helium-hydrocarbon gas mixture using the Cellulose Acetate (CA) and Silicone Rubber (SR) membranes. CA is more permeable to He and SR more permeable to hydrocarbon. Experimental data from a previous study (2) was used in model verification and as the basis for the analysis and the optimal design. The characteristics of the TMC used in the simulations are summarized in Table 1. Details are referred to Ref. (2).

RESULTS AND DISCUSSION

The accuracy of the mathematical model describing the separation of a binary gas mixture in a TMC was tested prior to simulation studies. Table 2 compares the computer simulation results with the previous experimental data on the separations of He-hydrocarbon mixtures (2). Excellent agreements between the predictions and the experimental results were achieved.

Table 1. Characteristics of the TMC

	CA Membrane Module		SR Membrane Module	
Number of hollow fibers	260		30	
OD/ID (μm)	290/150		634/310	
Length (cm)	46.0		149.0	
Total membrane area (m^2)	0.080		0.064	
Permeabilities[§]		α^*(He/a)		α^*(a/He)
He	199	1.0	35.6	1.0
CH_4	16.9	11.8	103.1	2.9
C_2H_4	15.3	13.0	271.2	7.6
C_2H_6	14.7	13.5	324.5	9.1

Temperature: 24°C
High pressure side (tube side): 2 atm.,
Low pressure side (shell side): 1 atm..

§ Unit: cm^3(STP)-cm/sec-cm^2-cmHg x 10^9
Values to the mixed gases differ from those to the pure gases for the CA membrane. Data refers to (2).

Table 2. Simulation results compared with experimental data

Run #	feed gas conc. (mol% He)	feed rate (cm^3/min)	top rate (cm^3/min) expt.	top rate calc.	top conc. (mol% He) expt.	top conc. calc.	bottom rate (cm^3/min) expt.	bottom rate calc.	bottom conc. (mol% He) expt.	bottom conc. calc.
1	He-CH_4	64.3	17.7	17.7	84.5	84.0	46.6	46.6	34.0	35.0
2	(48.5)	32.7	14.6	15.2	80.4	79.5	18.1	17.6	20.5	21.8
5		15.7	10.1	10.2	71.8	72.2	5.56	5.50	2.8	4.8
6	He-CH_4 (25.2)	23.1	5.50	5.54	67.0	67.7	16.6	17.6	10.1	11.8
10	He-C_2H_6	17.5	1.44	1.57	97.2	96.9	16.1	15.9	7.2	7.7
12	(15.7)	10.6	1.90	2.02	76.9	77.0	8.67	8.55	1.0	1.2
13	He-C_2H_4	42.1	8.30	8.36	87.5	87.5	33.8	33.7	19.9	21.7
15	(34.8)	11.3	3.90	3.91	90.6	89.5	7.43	7.43	2.5	6.0

An inspection of Table 2 reveals that most of the experimental conditions used in the previous study were not optimal. For example, run 10 yielded high top product purity (97 mol% He) but at low recovery (50% He), while run 5 provided higher recovery (95% He) but with lower top product purity (72 mol% He). Neither case was effective in terms of overall separation efficiency when moderate recovery (85-95%) and high product purity (>85 mol%) were desired. In the following sections, computer simulations were conducted to demonstrate the importance of optimization in improving the separation performance of a TMC.

Design of TMC in Terms of Top Product Purity

The inherent ability of the TMC to produce very high purity products by collecting the reject stream in each of its two stages contributes to the significant improvement in its separation performance over other designs under comparable conditions (2,3,4). Therefore, TMC is most suitable for separation problems with high product purity requirements. In the following simulations and analyses, the recovery of high purity helium from natural gas (He-hydrocarbon mixture) is considered. The optimization objective is to find the optimum membrane area ratio for a given total membrane area to maximize the He recovery while meeting the top product purity requirement.

Since stripping is more efficient than enrichment in producing a high purity product stream (2), increasing the SR membrane area in the second stage, in which the He product stream is withdrawn, would be more efficient than increasing the CA membrane area in the first stage to obtain the higher purity of He product. The separation of a He-CH_4 feed mixture containing 48.5 mol% He (run 5, Table 2) by a TMC under the operating conditions listed in Table 1 is used as an example to examine the effectiveness of the second stage of a TMC in improving its separation efficiency. The performances of TMC designs at the fixed feed rate per unit CA membrane area (M_{CA}= 0.08 m^2) and with increasing SR membrane area (M_{SR}) from zero to a maximum value are shown in Figure 2 in terms of product concentrations, membrane area M_{SR}, and the Rony's separation index (to be discussed later) as functions of stage cut θ_A (defined as the fraction of feed recovered as the He product stream). The top product concentration of a single CA membrane permeator (broken line) is also shown for comparison. A single membrane permeator with M_{CA}= 0.08 m^2 (the actual size of the CA membrane module in Table 1) would yield a permeate stream of 63.7 mol% He at θ=0.76. Increasing the SR membrane area to its actual size for experiment run 5 with M_{SR}=0.064 m^2 improves the top product purity to 72 mol% He, as shown by the experimental data point in Figure 2. As M_{SR} further increases, X_A approaches unity at a finite stage cut θ_A. For comparison, the maximum top product purity attainable with a single CA membrane permeator is 86.3 mol% at zero cut, while the same purity of the top product can be achieved by a TMC design in Figure 2 at a cut of 0.5.

The optimal performance of a TMC was achieved when two membrane areas were so designed that the maximum amount of He would be recovered at its high purity desired. In Figure 2, TMC with M_{SR}=0.7 m^2 yielded a top product with purity of X_A =99 mol% He at a cut θ_A=0.3. Further increase in M_{SR} beyond this point leads to little improvement in top product concentration X_A and great losses in recovery (cut θ_A) and the bottom product concentration X_B. For example, increasing the purity requirement from 99 to 99.5% would result in a substantial increase in membrane area requirement in the second stage. When M_{SR} increases to 1.33 m^2, θ_A approaches zero, indicating the occurrence of complete recycle in the second stage of a TMC. Therefore, the point at M_{SR}=0.7 m^2 represents the optimum value of M_{SR} for given M_{CA}. TMC so designed would maximize the recovery of high purity He product and therefore provide the optimum performance for the given separation problem.

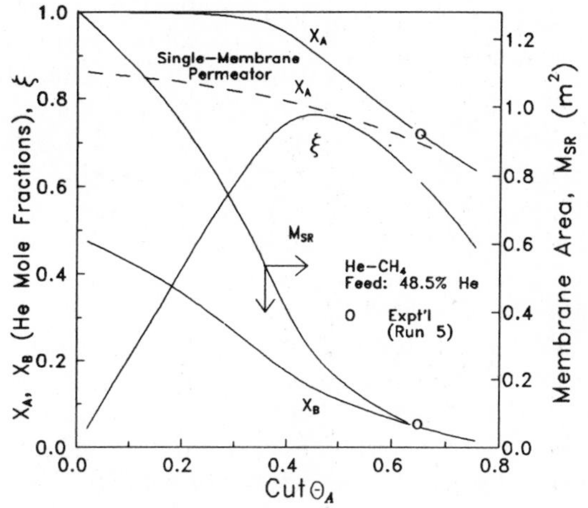

Figure 2. Separation of He-CH_4 mixture in a TMC with M_{CA}=0.08 m^2 and different Values of M_{SR}

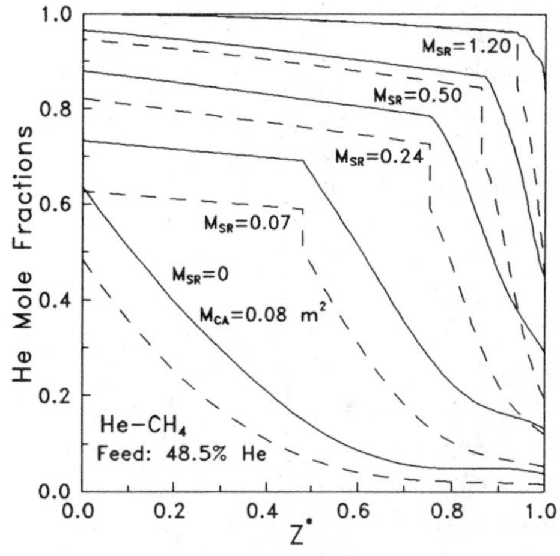

Figure 3. Concentration profiles in a TMC with M_{CA}=0.08 m^2 and different values of M_{SR} (m^2). —— permeate side, - - - feed side.

Figure 3 presents the concentration profiles in a TMC for several selected M_{SR} values examined in Figure 2. The abscissa is the normalized membrane area z^*. The point $Z^*=0$ corresponds to the top of a TMC and $Z^*=1$ the bottom. The feed location is indicated by the discontinuity in the feed-side concentration profile (dashed lines). As M_{SR} increased from zero to 1.2 m^2, concentration profiles were raised to higher levels by the more stripping actions of the SR membrane. The feed-side concentration profiles in the first stage were elevated by the recycling of the highly He-concentrated permeate stream from the second stage. The higher He concentration enhanced the driving force for permeation and therefore improved the overall separation efficiency of a TMC. Excessive increase in the membrane area of the second stage to $M_{SR} > 1.2$ m^2, however, would lead to great loss in the recovery of high purity top product due to its nearly complete recycle in the second stage of a TMC.

The most important design variable for a TMC is the membrane areas to be used in each stage of a column. The He-C_2H_4 separation under the feed conditions of run 13 was used to illustrate the effects of the total membrane area, M_t, and the area ratio of CA to SR membranes, R_m, on the TMC performance. Simulation results of the TMC designs with different values of M_t (1.4, 2.5, 3.5, and 4.5 m^2) and R_m (0.5, 1, 2, and 3) are presented in Figure 4 by plotting families of curves of the product concentration against the stage cut. The separation performance of a TMC at constant values of M_t, as shown by the solid curves, was determined by the CA/SR membrane area ratio. Higher He product concentration but lower stage cut was obtained with smaller value of R_m (meaning decreasing the first stage CA membrane area and increasing the second stage SR area). As the value of M_t increases, the concentration vs. cut curves shifted toward the direction of higher cut, approaching the maximum recovery limit. The separation performance of a TMC at constant values of R_m, shown by the relatively flat X_A vs. θ_A curves in broken lines, indicates that a TMC at a fixed value of R_m can obtain more product by increasing M_t with a slight variation in the top product purity. As the value of R_m decreases, X_A vs. θ_A curve moves up toward higher product purity and becomes fairly flat in a wide range of stage cut.

Significant improvement in top product purity of a TMC over that of a single-membrane permeator was achieved by properly selecting the membrane area ratio for a TMC design. For example, the top product purity of a TMC at $R_m=1$ is between 91 to 94%, while it approaches unity (99%$^+$ purity) at $R_m=0.5$ for any value of M_t. To meet the product requirement for

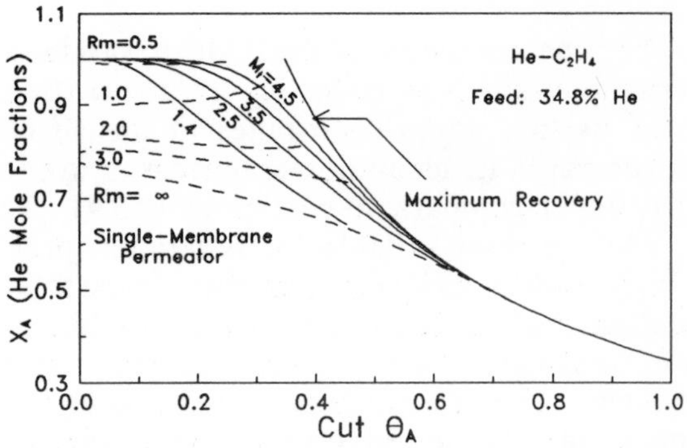

Figure 4. Separation of He-C_2H_4 mixture in a TMC with different total membrane area (m^2) and area ratio

higher than 90 mol% He purity, R_m must be smaller than 1. It is concluded, therefore, that the most important parameter for the TMC design is R_m. Results such as shown in Figure 4 can be used as an effective aid to the optimal design of a TMC. In actual design applications, the best approach to determine the amounts of two membranes in each stage is first to locate a working point in Figure 4 according to the top product purity (X_A) requirement and the recovery requirement (cut θ_A). The total membrane area and area ratio (M_t and R_m) can then be located by extrapolating from families of curves for constant M_t and constant R_m respectively. If the objective for the separation is to make the greatest possible amount of He product with set product purity, A constant M_t curve close to the maximum recovery limit should be chosen. Values of M_t and R_m can be determined from this curve according to the specified product purity. Such a TMC design would minimize the waste of membrane area and maximize the recovery of the desired species with high product purity requirement.

Design in Terms of Rony's Extent of Separation

From the results presented in Figures 2 and 4, it can be seen that increasing the membrane area M_{SR} and decreasing the membrane area M_{CA} would result in a purer top product, while decreasing M_{SR} and increasing M_{CA} would result in a purer bottom product. In both cases, the recovery of the end (top or bottom) product must be traded off for the higher product purity requirement. To achieve the optimum overall separation performance, compromise between the maximum purity and maximum recovery has to be made. Rony's extent of separation (16) can be used to provide such a measure of optimal separation. Definition of Rony's Index for extent of separation is given below in terms of stage cut and

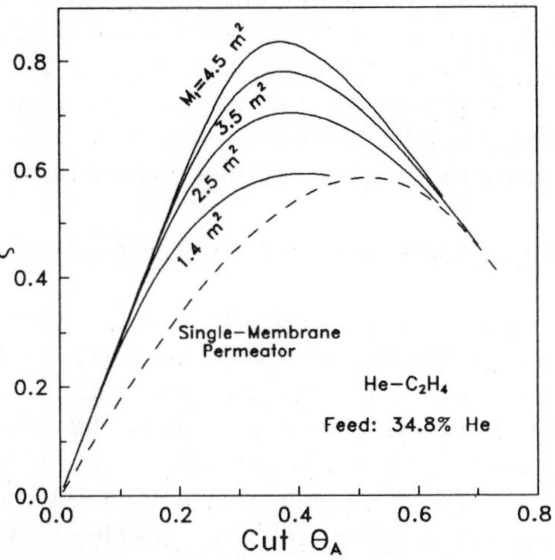

Figure 5. Extent of separation as a function of stage cut for a TMC with different total membrane area

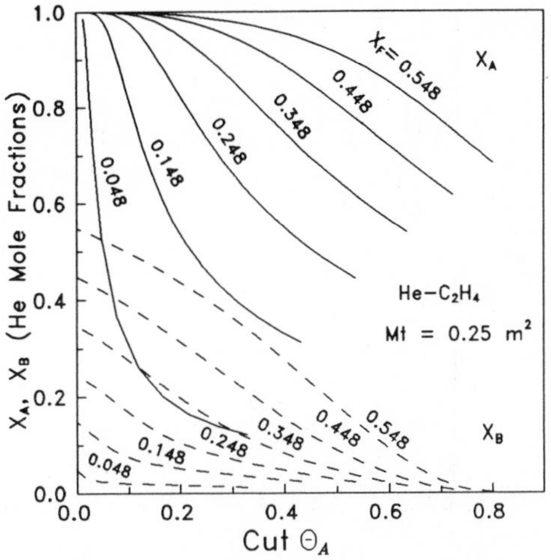

Figure 6. Feed composition effect on product concentrations in a TMC

compositions of the feed and the product streams:

$$\xi = \theta_A(X_A-X_F)/[X_F(1-X_F)]$$

This index lies between 0 and 1 when there are only two product streams for the binary system, with $\xi=1$ indicating perfect separation and $\xi=0$ no separation. The maximum value of Rony's index indicates the optimal separation of a binary gas mixture.

The Rony's index for the He-CH_4 separation was presented in Figure 2. The maximum separation index was reached at $\theta_A=0.485$. This condition corresponds to the point where top and bottom product purity vs. cut curves have approximately the same slope. Maximum extent of separation can be obtained through the best compromise between product purities of two end streams. The Rony's separation index for the He-C_2H_4 separation are presented in Figure 5. For each fixed total membrane area, the maximum ξ is reached at a specific value of θ_A. As the total membrane area increases, magnitude of Rony's index increases but the maxima always occurs at about the same cut. In addition, for all the cases examined, the maximum Rony's index invariably occurs at cut which has approximately the same numerical value as the feed composition. For example, for He-CH_4 separation (X_F= 0.485) in Figure 2, maximum ξ occurs at $\theta=0.485$ and for He-C_2H_4 separation (X_F=0.348) in Figure 5, maximum ξ occurs at $\theta=0.36$. In fact, if the total membrane area increases further, the maximum ξ would approach unity, indicating perfect separation for the given gas mixture. This means, by the material balance, that the amount of pure He recovered in the top product stream would always be equal to the value of X_F. This result further indicates that the TMC design based on the maximum Rony's index is an optimal one.

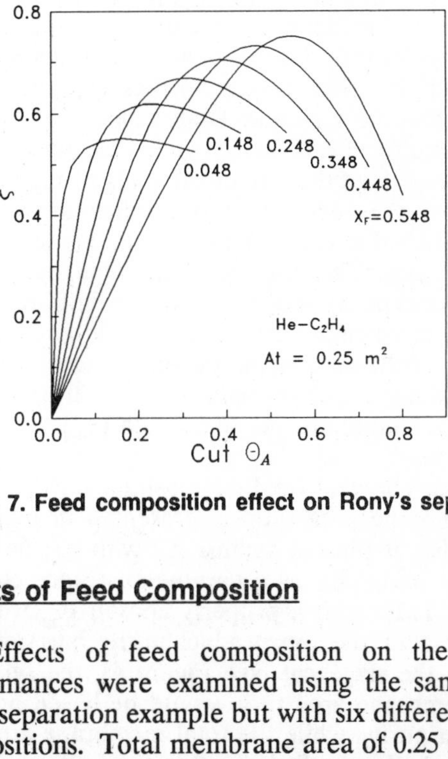

Figure 7. Feed composition effect on Rony's separation index

Effects of Feed Composition

Effects of feed composition on the TMC performances were examined using the same He-C_2H_4 separation example but with six different feed compositions. Total membrane area of 0.25 m^2 was used. In Figure 6, both top and bottom product concentrations are presented as functions of θ_A. All the X_A curves converge toward unity as cut θ_A approaches to zero while the X_B curves converge toward zero as θ_A approaches to the respective limiting values for each specific feed gas mixture. As indicated by the significant change in curves in Figure 6, both top and bottom product purities are very sensitive to the feed mixture composition. The curves are shifted towards higher concentration and higher cut with increasing feed composition.

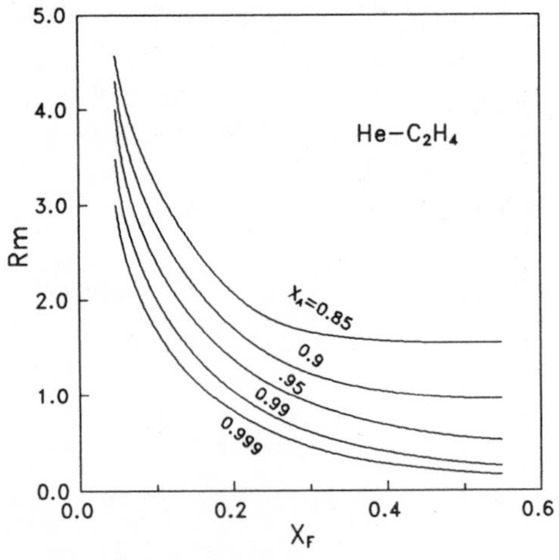

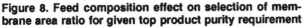

Figure 8. Feed composition effect on selection of membrane area ratio for given top product purity requirement

The Rony's index for the above cases is shown in Figure 7. The optimum cut at which the maximum Rony's index would occur is a strong function of feed composition. The results further support the conclusion that the maximum Rony's index for a particular feed mixture always occurs at a cut which approximately equals to the numerical value of feed composition. As the feed composition increases, ξ vs. θ_A curves shifted to the upper-right with higher maximum Rony's index. This, together with the higher attainable top product purity shown in Figure 6, indicates that the TMC is more competitive with conventional processes when the feed has a higher fast-gas concentration. This conclusion is consistent with the result from a previous study on membrane gas separation (17).

The effect of feed composition on the selection of membrane area ratio is presented in Figure 8, in which R_m is plotted against X_F with the desired top product purity X_A as a parameter. Since changes in M_t for TMC with a properly chosen R_m would only slightly alter the top product purity but significantly change the stage cut (see Figure 4), the selection of the membrane area ratio would be based on the top product purity, while the total membrane area on the stage cut. Figure 8 shows that for a specific X_A, R_m decreases with increasing feed composition. This is due to the fact that the higher driving force owing to a higher feed composition could reduce the requirement for the fraction of membrane area allocated to the first stage of a TMC. The sharp increase in R_m in the region of smaller X_F indicates that producing a highly purified helium product from a lean He feed mixture would require a significant increase in the membrane area in the first stage of a TMC. The large recycle stream of higher He concentration would offset the lower feed composition, providing an adequate driving force for a TMC. At a given feed mixture, membrane area ratio R_m decreases as the product purity requirement increases, indicating that a relatively larger membrane area in the second stage would be required to produce higher purity He.

Effects of Membrane Permselectivities

Another key factor affecting the TMC performance was the magnitudes of the two selectivities of the membrane pair. In the above simulation analysis based on the actual physical values, the selectivity of SR membrane toward C_2H_4 relative to He ($\alpha^*(C_2H_4/He)=7.6$) was not as high as that of CA membrane toward He relative to C_2H_4 ($\alpha^*(He/C_2H_4)=15.3$) (see Table 1). To examine the permselectivity effect on the TMC performance, a hypothetical separation problem is simulated with a hypothetical membrane to replace the actual SR membrane. The permeability of He through the hypothetical SR membrane was increased so that its selectivity was the same as that of CA membrane ($\alpha^*(C_2H_4/He)=15.3$).

Figure 9 compared the product concentrations and the Rony's extent of separation of a TMC using the actual SR membrane selectivity with a TMC using the hypothetical membrane with higher C_2H_4-selectivity. All other parameters are kept the same as in simulations for Figure 4. Both product purities and the overall separation efficiency of the TMC in terms of Rony's Index are improved significantly when a hypothetical membrane of higher selectivity is used. Using a more C_2H_4-selective membrane in the second stage improves not only the top product

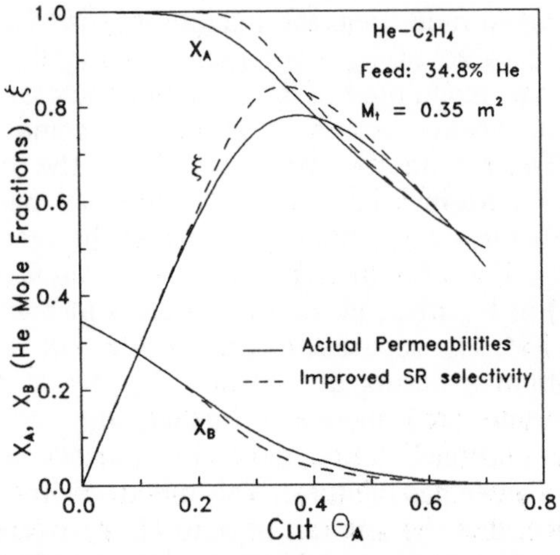

Figure 9. Effect of improved SR selectivity on TMC performance

purity but also the bottom product purity. This fact underlines the basic feature of the TMC that its two stages are coupled by the recycle of an intermediate-purity stream so that a change in the second membrane affects both end products. Consequently, the separation of a binary gas mixture by the TMC could be greatly enhanced by the use of two membranes with high reversed gas selectivities. Notice that the optimum cut at which the maximum ξ occurs still remains at the value of X_F, indicating that permselectivity of the membrane has little effect on the optimum cut for achieving the maximum extent of separation.

CONCLUSIONS

Two-Membrane Column (TMC) with properly designed membrane areas in its two stages for each specified feed condition provides an efficient separation both in terms of the high product purities and in terms of the overall separation efficiency. An increase in the relative membrane area in the second stage improves the top product purity. Excessive increase, however, leads to little improvement in top product purity and great losses in the top product recovery and bottom product purity. There exists an optimum membrane area ratio which maximizes the cut achievable with any given total membrane area in meeting the product purity requirement. On the other hand, the optimal design for achieving the best separation efficiency in terms of Rony's index assumes a different optimum cut which has a value approximately equal to the value of X_F regardless of total membrane area.

Membrane area allocation for each stage for the optimum performance of a TMC is a strong function of feed composition. Relatively larger membrane area in the second stage is needed for a richer feed mixture. Increase in the relative values of the two selectivities of the membrane pair improves the performances of the TMC.

ACKNOWLEGEMENT

The financial support from National Science Foundation (CBT 8810825) is gratefully acknowledged.

NOTATION

Roman letters

M_{SR} --- Silicone Rubber membrane area, m^2
M_{CA} --- Cellulose Acetate membrane area, m^2
M_t --- total membrane area, m^2
R_m --- membrane area ratio (CA membrane/SR membrane)
X_A --- top product concentration, mol%
X_B --- bottom product concentration, mol%
X_F --- feed composition, mol%
z^* --- fraction of total membrane area from the top to a point along the length of a TMC

Greek letters

α^* --- selectivity of membrane
ξ --- Rony's separation index
θ_A --- overall stage cut

LITERATURE CITED

1 Spillman, R.W., Chem. Eng. Prog., 41, (Jan. 1989)

2 Seok, D.R., S.G. Kang, and S.T. Hwang, J. Memb. Sci., 27, 1 (1986)

3 Yan, Zhiquan, and Y-K Kao, J. Memb. Sci., 42, 147, (1989)

4 Seok, D.R., S.G. Kang, and S.T. Hwang, J. Memb. Sci., 33, 71 (1987)

5 McCandless, F.P., J. Memb. Sci., 24, 15 (1985)

6 Hwang, S.T., and J.M. Thorman, AIChE J., 26, 558 (1980)

7 Sirkar, K.K., Separ. Sci. Technol., 15, 1091 (1980)

8 Sengupta, A., and K.K. Sirkar, J. Memb. Sci., 39, 61 (1988)

9 Stern, S.A., J.E. Perrin, and E.J. Naimon, J. Memb. Sci., 20, 25 (1984)

10 Perrin, J.E., and S.A. Stern, AIChE J., 32, 1889 (1986)

11 Perrin, J.E., and S.A. Stern, AIChE J., 31, 1167 (1985)

12 Blaisdell, C.T., and K. Kammermeyer, Chem. Eng. Sci., 28, 1249 (1973)

13 Walawender, W.P., and S.A. Stern, Sepa. Sci., 7, 553 (1972)

14 Villadsen, J., and M.L. Michelsen, Solution of differential equation models by polynomial approximation, Prentice-Hall (1978)

15 Kao, Y.K., and Zhiquan Yan, Chem. Eng. Comm., 59, 343 (1987)

16 Rony, P.R., AIChE Symposium Series, 120, 68, 89 (1972)

17 Goodin, C.S., Hydrocarbon Process, 61, 5, 125 (1982)

UPGRADING OF LANDFILL GAS BY MEMBRANES-PROCESS DESIGN AND COST EVALUATION

R. Rautenbach and H.E. Ehresmann ■ Institut für Verfahrenstechnik, Rheinisch-Westfälische Hochschule Aachen Turmstr. 46, D-5100 Aachen, Federal Republic of Germany

Gas permeation is a reliable process for upgrading biogas to natural gas pipeline specifications (90–95 vol-% CH_4). The paper reports four years of experience with a pilot plant where various module systems have been tested (Monsanto, Envirogenics, UBE Ind.). The influence of the main parameters on the process (gas temperature, -pressure, -composition, module arrangement) is discussed. Processing costs are compared to absorption and adsorption processes. In case of landfill gas, a special treatment step is required to remove H_2S, aromatics and halocarbons.

INTRODUCTION

In the digesters of sewage treatment plants and on landfill depositories, biogas is produced in large quantities. Biogas contains about 50 – 70 vol–% of methane, carbon dioxide, small amounts of water vapor and air. Furthermore, traces of hydrogen sulfide, halogenated hydrocarbons, aromatics and terpens can be contained in the gas.

Although biogas is an interesting energy source, its specific energy content is too low for a transport by pipeline even over short distances. There are two alternatives for an economical energy export:

- conversion into power by gas engines/generators on the site and feeding into the grid system
- methane enrichment on the site to about 90 96 vol–% and feeding the product into the natural gas distribution system.

The investigations reported here are related to the second alternative and confined to gas permeation using polymer membranes of the solution–diffusion type.

MATERIAL TRANSPORT IN GAS PERMEATION

In principle, gas mixtures can be fractionated by porous membranes and by pore–free membranes of the solution–diffusion type. The selectivity of porous membranes is, however, low compared to the selectivities which can be achieved with solution–diffusion type membranes. Accordingly, gas permeation gained interest after the introduction of pore–free polymer membranes, and even then only when they became available as asymmetric, high–flux membranes. The module types employed in gas permeation are of the hollow fibre and of the spiral–wound configuration.

The essential transport steps in such membranes are:

- sorption of the permeating component in the polymer
- diffusion of the permeating component through the polymer
- desorption of the component out of the permeate side (1).

The material transport of a permeating component is described with sufficient accuracy - at least for permanent gases like oxygen, nitrogen, and methane – by:

$$\dot{n}''_k = Q_k\,(p_F\,x_k - p_P\,y_k) = Q_k\,p_F\,(x_k - \frac{p_P}{p_F}\,y_k). \quad (1)$$

According to equation 1, the molar flux

$\dot{n}''_k$ of any permeating component k is proportional to the difference of its partial pressures across the membrane. The constant Q_k, the permeability of the membrane for the component k, can be determined from permeation experiments with the membrane. For design and simulation of membrane modules and processes, two parameters are important:

- the ideal separation factor $\alpha_{ij} \equiv \frac{Q_i}{Q_j}$ which follows from the general definition of the selectivity and equation 1 for $p_P \to 0$
- the pressure ratio $\frac{p_F}{p_P} \equiv \delta$

Figure 1 shows the influence of these parameters on the permeate composition y for different feed compositions x. It clearly indicates that for low concentrations of the preferentially permeating component in the feed, a pure permeate cannot be obtained in a one-stage process, even for very selective membranes and very high pressure ratios.

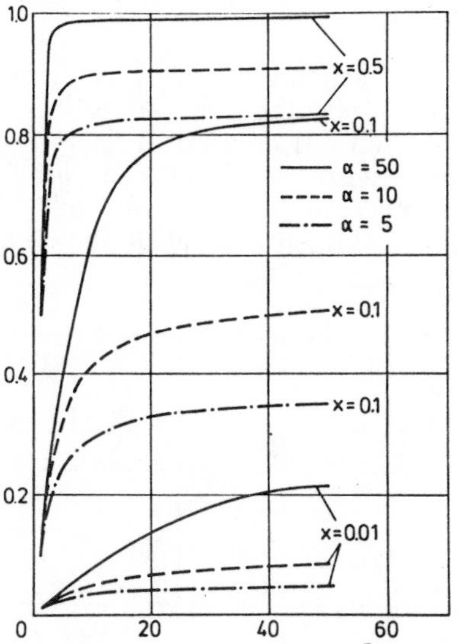

Figure 1. Influence of pressure ratio, ideal separation factor and feed composition on permeate composition

In principle, the pressure ratio can be increased by either raising the feed side pressure or by reducing the permeate side pressure (applying vacuum). However, an analysis shows that the application of vacuum to the permeate side is only reasonable in cases where the permeate side pressure losses due to friction are kept very low by adaequate module design.

It must be emphasized that equation 1 has to be applied locally. As a result of flux across the membrane, the relevant parameters like flow rates at the feed and permeate side of the membrane, gas compositions and pressures will vary along the membrane. This has to be taken into account in all module and process calculations.

EXPERIMENTS

Since 1985 we operate a pilot plant for methane enrichment of biogas by gas permeation on the premises of a municipal sewage treatment plant. The flexible plant design with feed flow rates up to 70 m^3(STP)/h, feed pressures up to 60 bars and maximum temperatures of 130 °C allows the installation of different module systems (**Figure 2**). Three systems, Envirogenics spiral-wound module equipped with cellulose acetate membranes and the hollow fibre modules of Monsanto Co. and UBE Ind. with

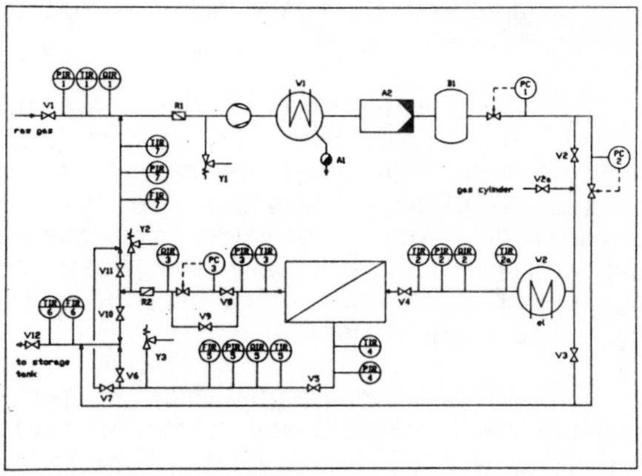

Figure 2. Flow sheet of the gas permeation pilot plant

polysulfone/silicon resp. polyimide membranes have been tested for several thousand hours each. Based on the results of this project, a gaspermeation plant with a capacity of 200 m^3-(STP)/h is currently under construction on the premises of a landfill depository; start up is scheduled for May 89. In order to meet pipeline gas specifications, this plant includes a pretreatment stage for the removal of the trace components H_2S and halogenated hydrocarbons (product: H_2S < 1 ppm, Cl,F < 5 ppm).

Influence of Operating Parameters on the Process

With gas permeation, the separation effect

occurs due to the differences in transport velocity of the individual gas components through the membrane, commonly refered to as "slow" and "fast" gases. In general, all polymers are qualitatively similar with respect to selectivity: they are highly permeable for the biogas components CO_2, H_2O, H_2S and O_2 and less permeable for nitrogen and methane. Consequently, the product of sewage or landfill gas, the methane, is obtained at the retentate side. Because nitrogen and methane are transported through the membrane at similar velocities, the maximum achievable methane content is limited by the nitrogen concentration of the raw gas. In the following, the operating behaviour of a gas permeation plant for methane enrichment will be discussed. Of major interest is the influence of process parameters such as feed gas flow rate, feed gas pressure, feed gas temperature on product composition and recovery rate. If not specified, the standard feed gas composition for the experiments was appr. 67 vol–% CH_4, 32 vol–% CO_2 and 1 vol–% air; water vapor had been removed from the gas by a pressure swing adsorption dryer after compression to a dew point at operating pressure of about –25 ^{0}C.

Beside the influence of the above mentioned process parameters on product quality and recovery rate, the influence of different module arrangements like one and two stage cascades has been investigated. The Monsanto unit consisted of four and the Envirogenics modules of 15 elements which could be operated in various configurations.

Figure 3 shows the influence of operating pressure (feed pressure) and membrane area on product purity and recovery rate. According to **figure 3**, an increase of the pressure ratio $p_{F\alpha}$ / $p_{P\omega}$ results in a higher purity of the slower permeating component methane with the disadvantage of reduced recovery rates. With increasing pressure ratio, the feed gas is faster depleted of carbon dioxide. Consequently, the methane partial pressure increases which in turn leads to higher methane flux in the last section of the membrane stage.

The influence of feed gas temperature on methane purity and recovery rate is shown in **figure 4**. For a constant feed gas pressure of 32 bars and flow rates of 10 and 20 m^3(STP)/h, temperature has been varied between 30 ^{0}C and 80 ^{0}C. According to **figure 4**, an increase in feed gas temperature leads to higher methane purities at the cost of reduced methane yield. Operation temperature of the modules is important, since generally the flux increases with temperature obeying an Arrhenius type law, while selectivity decreases. It has to be emphasized that **figure 4** shows the influence of the feed entrance temperature on unit performance and not the influence of local temperature on

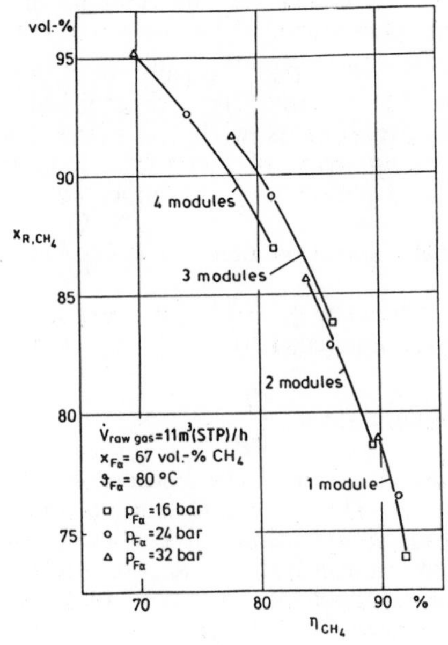

Figure 3. Influence of feed pressure and membrane surface area on product purity and recovery rate

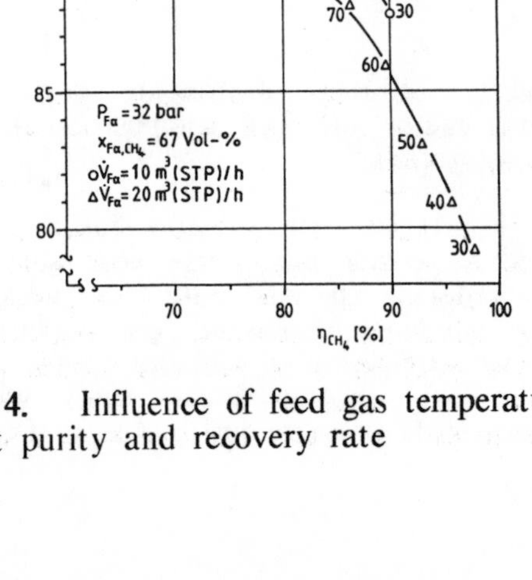

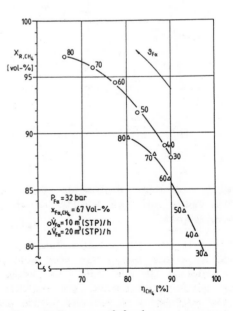

Figure 4. Influence of feed gas temperature on product purity and recovery rate

the local separation characteristic. Because of the integral character of **figure 4**, temperature drops resulting from the Joule–Thomson effect are implicitly contained in these curves. Carbon dioxide and methane exhibit a pronounced Joule–Thomson effect, i.e. a temperature drop in case of isenthalpic expansion. This effect was unimportant in the early applications of gas permeation (hydrogen / nitrogen separation), but has to be taken into account in case of biogas separation.

MODULE CONFIGURATION

All results discussed so far have been obtained with a one stage process where all modules were connected in series. The experiments demonstrate that in cases, where the retentate represents the product, any desired product purity can be achieved for a binary system regardless recovery rate. However, the recovery rate can be raised by a two stage reflux cascade where the permeate of the second stage is recycled to the compressor inlet. Here, an additional compressor is not necessary, but the flow rate is considerably increased compared to the one stage operation. Some results of our experiments are shown in **figure 5**, where a one stage and a two stage reflux process of identical total membrane surface area are compared. According

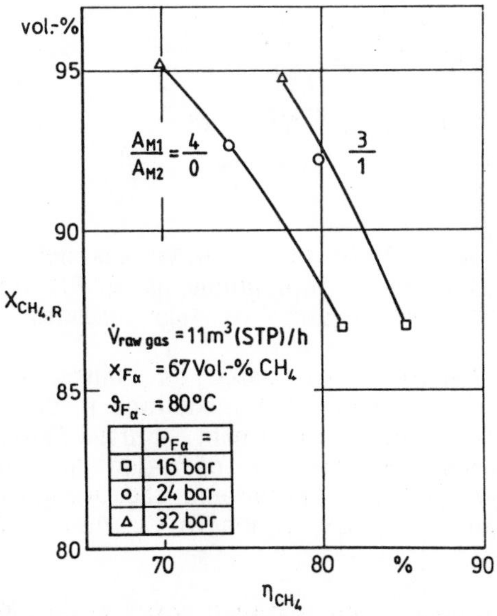

Figure 5. One and two stage cascade with identical total membrane surface area

to **figure 5**, the two stage reflux cascade shows significantly higher methane recovery rates.

MATHEMATICAL MODELING OF THE PROCESS

Gas permeation is one of the membrane processes where a detailed mathematical modeling is possible and reasonable, since such difficult effects like gel-layer formation, fouling or even concentration polarisation usually do not exist. Based on equation 1 for the local material transport and on permeation experiments with one module element and gas mixtures of different compositions, we developed a mathematical model (2) of the process accounting for:

- the flow pattern in the individual module, i.e. cocurrent, countercurrent or cross flow
- the multicomponent nature of the gas mixture
- friction losses at the permeate side of the membranes
- the Joule–Thomson effect.

Friction losses at the feed side (high pressure side) and axial backmixing due to diffusion have been neglected.

In **figure 6**, some results calculated for the one stage process consisting of four modules connected in series are compared with experimental results. According to **figure 6**, the calculations are in good agreement with the experiments. Attention should be given to the fact that the separation characteristic of the unit

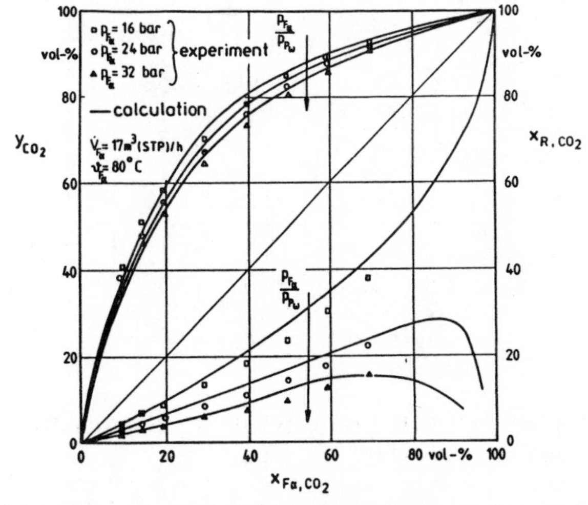

Figure 6. Separation characteristic of gaspermeation process – comparison between experimental and calculated data

actually can decrease with increasing pressure

ratio as far as the permeate is concerned. This is contrary to the well known local selectivity characteristic which must increase with increasing pressure ratio.

GAS PRETREATMENT – SEPARATION OF TRACE COMPONENTS

The long time operation of our pilot plant has proved that gas permeation with the presently available membranes is a reliable process for upgrading of biogas to the caloric specifications required by the gas distributers. Besides the caloric specifications, however, specifications regarding upper limits of hydrogen sulfide ($\leq$ 1 ppm) and halogenated hydrocarbons ($Cl_2 \leq$ 5 ppm) must be met; these very low figures cannot be achieved by gas permeation. For this reason, our new plant includes a two stage activated carbon adsorption prior to compression (**figure 7**). In the first stage, hydrogen sulfide is catalytically and partially oxygenated to elemental sulfur which is adsorbed in the pores of the activated hydrocarbon. The sulfur–loaded carbon, containing about 1 kg sulfur per 1 kg carbon, is then deposited or incinerated. The second stage contains activated carbon, specially treated for the adsorption of halogenated hydrocarbons. Other trace components like aromatic hydrocarbons and terpens are also removed in this stage (3).

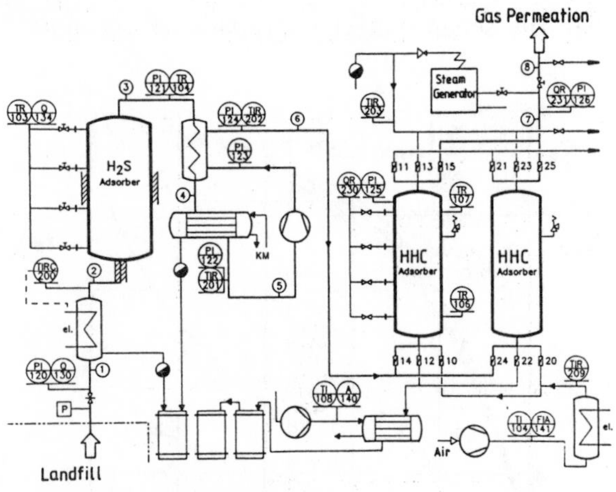

Figure 7. Gas pretreatment for the removal of trace components

The second stage is designed as a thermal–swing adsorption, regenerated with steam. The trace components are recovered in form of an aqueous mixture which is further treated by well known conventional methods. With about 25 % of the total investment costs of such a plant, the two stage pretreatment is an important factor regarding the economics of upgrading biogas.

COSTS AND PROFITS

Figure 8 is the result of a detailed optimisation calculation for a client. In this case which is to a large extent determined by the cost of power and the guaranteed price of the product, a two stage reflux cascade with about equal membrane surface area in both stages proved to be best.

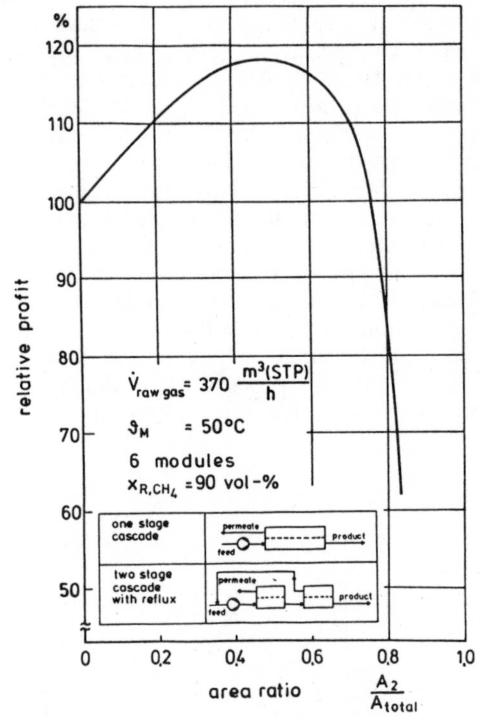

Figure 8. Profit of optimized gas permeation plants for biogas upgrading as a function of area ratio for one and two stage cascades

The calculations are based on contractor bids for the compressors, the membrane modules and the two stage pretreatment. Furthermore, the calculation accounts for operation and maintainance, instrumentation and the necessary installations for a connection to the municipal gas network.

Besides such detailed calculations for a special situation, more general calculations comparing gas permeation (GP) with alternatives are of interest, i.e. with:

- physical absorption using water (H_2O)

- chemical absorption using monoethanolamin (MEA)
- pressure swing adsorption using carbon molecular sieve (PSA).

Such calculations can and should be defined to the methane - carbon dioxide separation step, since the cost of pretreatment, operation and maintenance and interconnection are about the same for all processes. The results of a comparison between gas permeation and the alternatives is shown in **figure 9** (4).

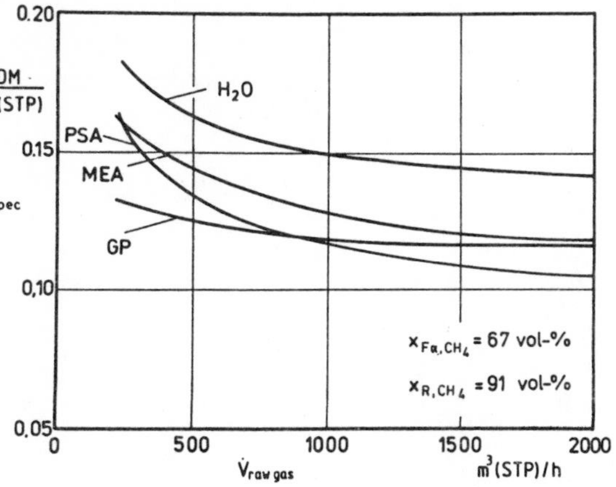

Figure 9. Comparison of specific product cost between gas permeation and alternative processes

According to **figure 9** where the specific separation costs are plotted against plant capacity, gas permeation is superior to all alternatives in the capacity range below 900 m^3(STP)/h. Since membrane costs are the major part of the total investment costs of a gas permeation, and since gas permeation is a relatively new process, further cost reductions for gas permeation can be expected in the near future.

CONCLUSIONS

Biogas, consisting mainly of methane and carbon dioxide, is produced in fairly large quantities (up to 5000 m^3(STP)/h) in sewage treatment plants and on landfill depositories. Carbon dioxide can be effectively separated from the biogas by gas permeation, resulting in a product containing 90 % methane or more. Gas permeation is a simple process and, according to our own four years experiments, highly reliable.In order to meet the specifications of the distributors, certain trace components have to be removed down to very low levels. This requires a gas treatment especially for the separation of hydrogen sulfur and, in case of landfill biogas, the separation of halogenated hydrocarbons, aromatics and terpenes. Pretreatment by a two stage adsorption using activated carbon, the presently available technique, is relatively expensive: about 25 % of the total investment costs have to be allocated to this pretreatment. The optimal module configuration - "one stage" or "two stage reflux" - has to be evaluated individually since it depends on site-specific factors and on factors like power costs and guaranteed product price. Compared to alternatives, gas permeation seems to be superior to all absorption processes, independend of capacity. Compared to pressure swing adsorption it is superior at least up to capacities of 900 m^3(STP)/h.

NOTATION

Roman Letters

$\dot{n}''$ = permeate flux [mol/(m^2·s)]

p = pressure [bar]

Q = permeability [mol/(m^2·h·bar)]

$\dot{V}$ = flow rate [m^3(STP)/h]

x,y = mole fraction [-]

Greek Letters

α = ideal separation factor
δ = pressure ratio
η = yield factor
Θ = cut rate

Subscripts and Superscripts

F = feed
i,j,k = gas component
P = permeate
R = retentate
α = module entrance
ω = module exit

LITERATURE CITED

1. S.T.Hwang, K.Kammermeyer, "Membranes in Separation", John Wiley & Sons, New York (1975)

2. R.Rautenbach, W.Dahm, "The separation of multicomponent mixtures by gas permeation", Chem.Eng.Process., 19 (1985)

3. K.D.Henning, E.Richter, K.Knoblauch, H.Jüntgen, "Reinigung und Weiterverarbei-

tung von Deponiegas mit Adsorptionsverfahren", Recycling International, 1984

4. R.Rautenbach, H.E.Ehresmann, H.Weyer, "CH_4-Gewinnung aus Faulgas durch Membrantechnik",Int. Recycling Kongress, Berlin, FRG (1986)

DESIGNING A TWO-STAGE RECYCLE MEMBRANE PERMEATOR CASCADE FOR MULTICOMPONENT GAS SEPARATIONS

Shihan Chen and Yuen-Koh Kao ■ Department of Chemical and Nuclear Engineering, University of Cincinnati, Cincinnati, OH 45221

Characteristics of a recycle permeator and its cascade configurations for multicomponent gas separations were explored using a mathematical model. The separation of CH_4-CO_2-N_2 mixture was used as an example to show how the model can be employed for designing membrane permeation systems. The design strategies for separating the most permeable species or the species with an intermediate permeability were discussed in the paper. It was found that a two-stage recycle permeator cascade was more effective than a single-stage design for most practical separation problems. For the separation of the most permeable species from a ternary mixture, the optimal two-stage cascade design can be practically achieved when both stages were designed on the basis of equal recovery and equal separation factor.

The development of highly selective, energy-efficient, nondestructive, and economical separation processes can bring significant economic benefits to many chemical processes. Among numerous novel processes, membrane permeation, due to its energy efficiency and simplicity, is becoming more attractive for gas separation applications. Industrial examples include Du Pont's Permasep (1, 2), Dow's Geron air separation system (3) and Monsanto's Prism Separator (4 - 6). These processes have been developed for the separation of hydrogen, carbon monoxide, carbon dioxide, ammonia and other industrial gases and for the sweetening of natural gas.

Extensive research effort has been made to investigate the engineering principles of membrane-based separation processes (7 - 9). Permeator models have been developed to simulate the performance (10 - 14). However, most publications deal only with binary gas systems even though many practical processes involve multicomponent mixtures. Research on multicomponent separations has been largely limited to experimental investigations (15 - 17). Very little has been said about design, whose improvement certainly contributes to the competitiveness of membrane processes against other separation processes.

The primary concerns with the design of a practical membrane separation system are membrane areas and power consumption, the former reflecting the investment capital and the latter dictating the operating cost. In a good design, which minimizes both the membrane area and power consumption, one should choose the best configuration in addition to individual design parameters such as pressure ratio between the feed and permeate streams, recycle fraction for a recycle permeator, and relative membrane areas for a cascade system. Many types of permeator configurations have been discussed in the literature. While one type may be better than another for a given separation problem, we feel it is important to understand the underlying design principles. For this purpose, a countercurrent capillary permeator with recycle and its cascade configurations (Figure 1) are selected as the focus of discussion in this paper.

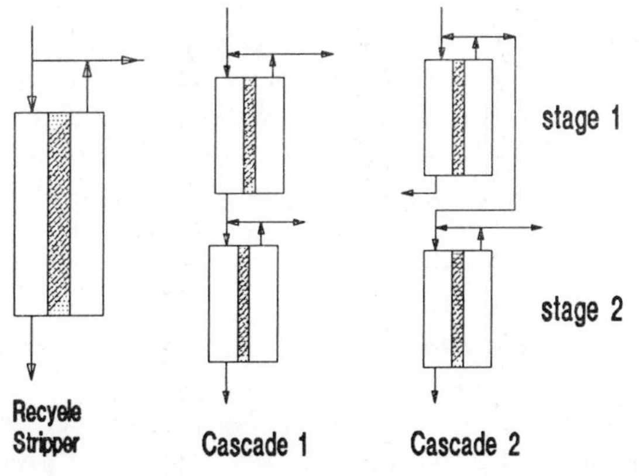

Figure 1. Recycle Permeator and Recycle Cascades

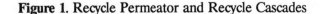

Correspondence should be addressed to Yuen-Koh Kao.

The separation of CO_2-CH_4-N_2 mixture with silicone rubber (SR) hollow fibers and/or cellulose acetate (CA) membranes is used as the example to illustrate the design procedures. The conclusions obtained in the following discussion could be readily extended to systems with more than three components. The justification of using this particular mixture can be found elsewhere (18). The permeabilities of pure gases in these membranes are listed in Table 1. The feed compositions are assumed to be 0.3, 0.4 and 0.3 for CO_2, CH_4 and N_2, respectively.

Table 1. Pure Gas Permeabilities*
(10^{-14} mol m/m^2 atm sec)

Gas Membrane	CO_2	CH_4	N_2
SR	92.7	26.2	7.91
CA	31.5	38.0	1.33

* Sources: (9, 17, 18)

A mathematical model of countercurrent capillary permeators for multicomponent mixtures incorporating the axial diffusion effects was used for the present study. The model consists of a set of ordinary differential equations and boundary conditions specified at both ends of the permeator. The model equations were solved by the orthogonal collocation method. The main advantage of using this numerical method is that no iteration is necessary regardless the type of operating conditions specified. The numerical algorithm and accuracy have been discussed elsewhere (19).

CHARACTERISTICS OF MULTICOMPONENT SEPARATIONS

Although the effects of design parameters on permeator performance have been extensively studied for binary separations, the results may not be directly applicable to multicomponent systems. It is important, therefore, to extend these results to multicomponent separations. The separation of the CO_2-CH_4-N_2 ternary mixture with the SR membrane will be used in the following discussion.

It has been established that in the separation of a binary mixture by a membrane permeator without recycle (simple stripper), an increase in the membrane area raises the recovery, R (the fraction of product species recovered from the feed), but lowers the product concentration of the more permeable species. The product concentrations achievable in a simple stripper are limited by the feed/pressure ratio and permselectivity. In order to attain concentrations and recoveries beyond the separation capacity of a simple stripper, a recycle permeator must be used. It is also known that there is a practical limit in the product purity achievable by a recycle permeator. Beyond this limit, a significant increase in membrane area leads only to a slight improvement in the product purity while maintaining the same recovery.

The separation of the most permeable species, CO_2, from the CO_2-CH_4-N_2 ternary mixture exhibits similar behavior as described above. This is summarized in Figure 2.

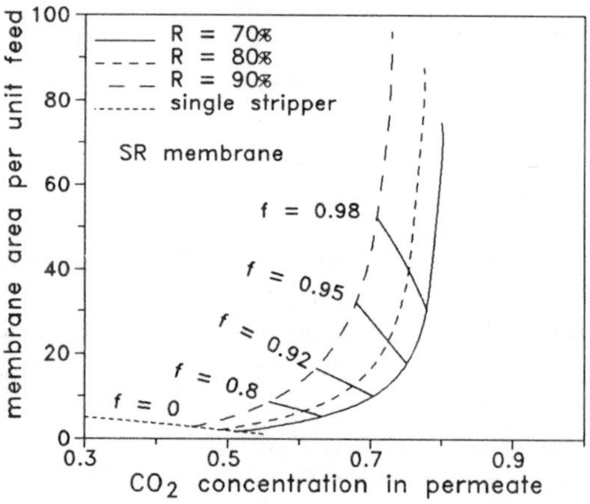

Figure 2. Separation Behavior of Recycle Permeators at Constant Recoveries

Figure 2 shows the specific membrane areas (area/unit of feed) of a recycle permeator as a function of CO_2 permeate concentration at three fixed recoveries (70, 80, and 90%) and of a simple stripper. For a simple stripper, there is a unique recovery for each achievable CO_2 concentration, as illustrated by the fine dashed curve. A stripper with a larger area offers a higher recovery at a lower CO_2 concentration. For the three recoveries specified, the permeate concentrations of the simple stripper are represented by the three points where the respective recycle permeator curves meet the simple stripper curve. Higher CO_2 product concentration must be obtained by employing recycle. As in binary separations, larger membrane areas along with higher recycle fractions, f (defined as the fraction of permeate used as recycle stream), produce higher CO_2 concentration products at a constant recovery. When the recycle fraction exceeds a certain range (0.95, for example), a substantial increase in membrane area can only increase the product purity slightly. The primary cause for the sharp increase in membrane areas is the excessive recycle action. For example, almost 95% of the permeate stream must be recycled to achieve a CO_2 composition of 0.7 at 80% recovery. Therefore, there exists a product concentration range beyond which any further enrichment will become very difficult and economically unfavorable. Although it is theoretically possible to fully recover the most permeable component to 100% concentration with a single recycle permeator, a cascade configuration would be a better alternative design. This point will be elaborated on later.

The permselectivity and the feed/permeate pressure ratio are two important parameters affecting the performance of a recycle permeator. The effects of these two parameters on ternary separations are shown in Figure 3 for a fixed recovery of 0.9. The specific membrane area needed for achieving the same permeate concentration is considerably reduced by doubling the feed/permeate pressure ratio. The practical upper limit of CO_2 concentration is also higher. The pressure effect is more significant in the high CO_2 concentration range. Because a moderate increase in the feed pressure is relatively easy and does not significantly affect the operating cost, many industrial permeation systems use much higher pressure ratios (20). The membrane selectivity has a more profound effect on the performance. The membrane area requirement is drastically reduced and a much higher CO_2 concentration is obtained if the permeability of the most permeable species is doubled. The power consumption has a similar dependency on the permeability and on the pressure ratio since larger membrane areas require more recycle action which then requires larger power. It should be recognized, however, that the availability for alternative membranes is rather limited at the present time.

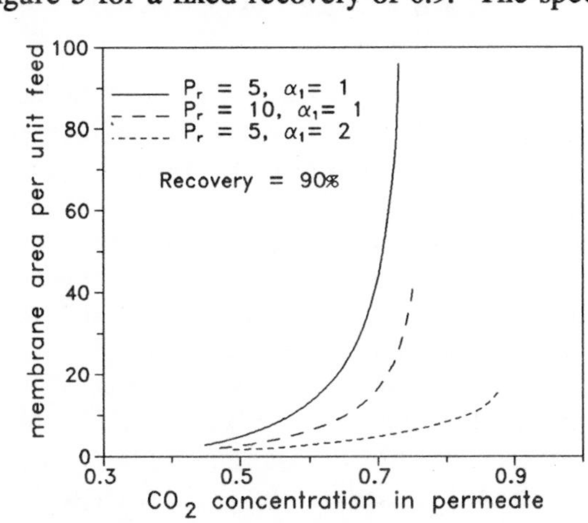

Figure 3. Effects of Feed/Permeate Pressure Ratio and Permselectivity on the Performance of a Recycle Permeator

While the separation of the most permeable species from a multicomponent mixture is similar to binary separations in general, the above conclusions are no longer valid for the recovery of a species with an intermediate permeability. Depending on its composition in the feed mixture and the permeabilities of all species, the intermediate species could be concentrated either in the permeate stream or in the reject stream. However, it is very difficult to highly concentrate the intermediate permeable species in either stream. Figure 4 shows the effects of the presence of CO_2 as an impurity on the separation of CH_4-N_2 at a fixed recycle fraction of 0.3. In this figure, the CH_4 feed concentration is 40%, CO_2 ranges from 0 to 15% and N_2 varies accordingly. In the absence of CO_2, higher CH_4 concentration can be obtained by lowering the cut. The highest CH_4 concentration appears at zero cut. When CO_2 is present as an impurity, the situation becomes different. The presence of CO_2 in the feed mixture lowers the achievable CH_4 concentration. The decrease in achievable CH_4 concentration is always larger than the CO_2 concentration in the feed. This is due to the higher permeability of CO_2 which tends to become more concentrated in the permeate stream. Moreover, the maximum CH_4 concentration occurs at an intermediate cut when the CO_2 concentration exceeds a certain level. Reducing the cut can no longer guarantee higher permeate concentrations. This is

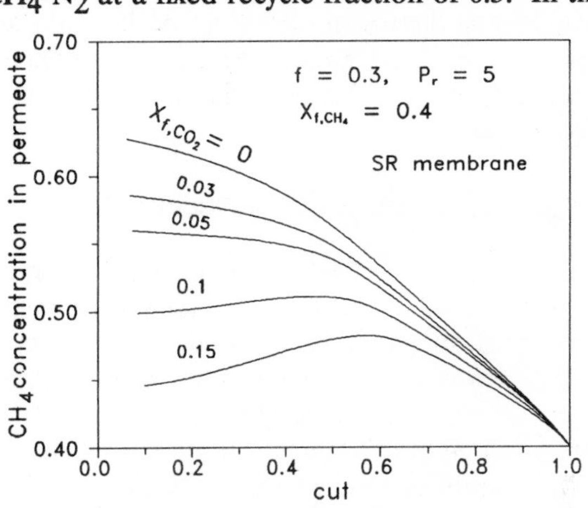

Figure 4. Effects of Impurity Having the Highest Permeability on Recycle Permeators

a unique situation for multicomponent separations.

DESIGN CONSIDERATIONS

In most practical design problems, in addition to the feed composition which is usually predetermined by the upstream conditions, the product purity and recovery of the desired species are usually specified. The design objective is to meet these requirements while minimizing the specific membrane area and power consumption. The design variables include the choice of membranes, the feed/permeate pressure ratio and the membrane areas of each module when a cascade arrangement is used. The effects of first two parameters on the separation performance have been discussed in the previous section. For practical separation problems (high recovery and high concentration requirements) a cascade arrangement is frequently needed. Therefore, the following discussion will focus on how to design a two-stage recycle permeator cascade. The conclusions obtained here could be extended to a design of multi-stage cascade.

Recovery of the Most Permeable Species

As shown earlier in Figure 2, an enormous membrane area is needed for a recycle permeator to meet design specifications located in the region where the membrane area-concentration curve becomes very steep. A single-stage design will be very inefficient under such a circumstance. An effective solution is to use a cascade configuration. Cascade 2 in Figure 1 should be used in this case because the objective is to separate the most permeable species. The design strategy for the cascade is simple. The permeate stream from the first stage, which will be the feed to the second stage, should be designed for a recovery higher, albeit at a composition lower than the design specifications. The second stage then further separates this intermediate product to meet the design specifications.

The design strategy is shown graphically in Figure 5. Suppose that the design specifications for the product concentration and recovery are y_p and r, respectively. The single-stage design can meet the specifications with a membrane area A. In a two-stage design, if the first stage is to enrich the most permeable species to y_1 at a recovery $r_1 > r$, the second stage must then further enrich the most permeable species to y_p at a recovery of r_2 such that $r_1 \cdot r_2 = r$. Thus the overall recovery and the final concentration will meet the design specifications. The required areas are A_1 and A_2-A_1 for the first and second stages, respectively, with a total area of A_2. Obviously, the two-stage design requires much smaller area than the single-stage design.

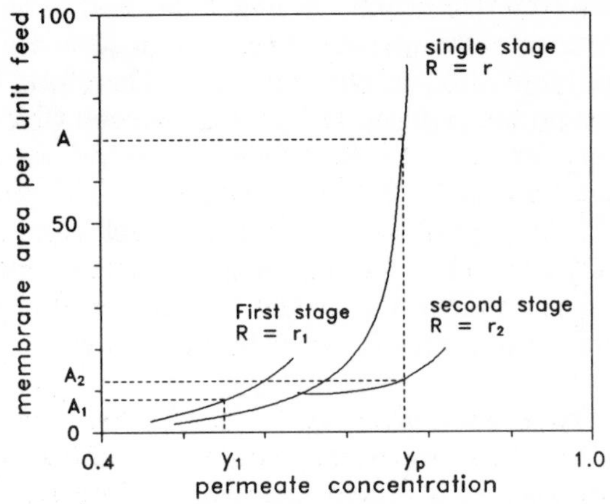

Figure 5. The Design Procedure for Two-Stage Recycle Cascade Systems

For numerical illustration, we assume that the design specification is to recover 80% of CO_2 in the feed to a concentration of 85%. For this particular design problem, the two-stage cascade must be used. It would be very expensive (practically impossible) to meet the design specifications with a single-stage design, as indicated by the solid curve in Figure 6. The total membrane area can be significantly reduced by using an additional stage. Consider a two-staged cascade in which the first stage recovers 89.5% of CO_2 to a composition of 61%, as represented by point B on the first-stage curve; the second stage further concentrates the product from the first stage at a recovery of 89.5%, along the dashed curve b, to the mole fraction of 0.85. The overall recovery will still be 80% and the end product meets the design specifications. In this design, the total membrane area requirement for the two stages will be less than 20 m^2/(mol/s).

While the above discussion clearly demonstrates that by using a two-stage cascade configuration one can substantially reduce the amount of required membrane area, it still leaves two important questions unanswered: how to choose the areas of each stage and at what CO_2 concentration should the permeate stream leave the first stage. Figure 6 shows three different designs as represented by A(a), B(b) and C(c) (A, B and C represent the permeate concentrations leaving the first stage, a, b and c are the corresponding second stage curves of the three respective designs.). All three designs meet the same design specifications. However, the total membrane areas increases as the designed permeate concentration of the first stage either increases from 61% (B) to 66% (A) or decreases to 56% (C). Clearly, B(b) represents the best design. While the exact solution could be obtained by optimization, a rule of thumb that can provide a near optimal design can be deduced from the above observation.

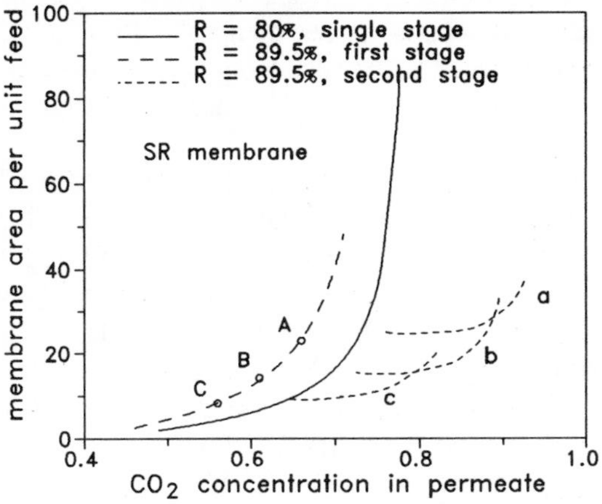

Figure 6. Effects of First-Stage Design on Total Membrane Area Requirement

One such rule for designing a two-stage cascade is to require each stage for the same recovery. For an overall recovery of 80%, each stage should achieve 89.5%. Another rule is that both stages achieve the same separation factor. Referring to Figure 5, this means:

$$\frac{y_1(1 - x_f)}{(1 - y_1)x_f} = \frac{y_p(1 - y_1)}{(1 - y_p)y_1}$$

where x_f is the feed composition. In this design problem, the CO_2 feed composition is 0.3 and the product concentration is specified to be 0.85. Thus the permeate CO_2 concentration leaving the first stage should be 0.61. As evidenced by the similarity between the curves of two stages in Figure 6, the shapes of the constant-recovery area-concentration curves are basically the same for different feed concentrations. Therefore, one can reasonably assume approximately the same separation characteristics for both stages. With this premise, the rational behind these rules becomes simple: at higher separation factors and higher recoveries, the extra membrane area required for achieving the same amount of concentration advancement (separation factor) or additional recovery is larger than that required at lower separation factors and lower recoveries. This is a typical "diminishing return" phenomenon. In order for both stages to operate at the highest efficiency, the separation duties should be equally shared by the two stages.

Although it is difficult to prove the above rules theoretically, the positive results obtained using these criteria justify their usefulness. Furthermore, as shown in Figure 6, the slope of the area vs. concentration curve, which is inversely related to the separation efficiency, of the first stage at $y_1=0.61$ (point B) is roughly the same as that at $y_p=0.85$ for the second stage. The slope mathematically represents the area increment needed for achieving a unit increase in concentration. An optimal design should be one in which the total membrane area is the lowest. The only way to achieve this is by requiring that the slopes of all stages be the same. Indeed, by moving the permeate concentration of the first stage (or the feed concentration to the second stage) away from 0.61, the slopes will no longer be equal and the design becomes less than optimal.

Recovery of the Species with an Intermediate Permeability

Although not a desirable situation, it does occur when the feed mixture contains impurity having the highest permeability. A typical example is the separation of CH_4 from the CH_4-N_2 mixture containing CO_2. In this example, the CH_4 has an intermediate permeability in the SR membrane. One way to avoid the separation of intermediate permeable species obviously is to choose a different membrane in which the desired species is most permeable. Then the design procedure discussed in the previous section can be applied. Such an alternative membrane, however, does not always exist. In some cases, one may find an alternative membrane but the permselectivities between the desired species and others are too small for the separation to be practical. For example, the selectivity between CH_4 and CO_2 is very small in the CA membrane although CH_4 is more permeable. Under the above circumstances, it is impossible to accomplish the given separation task by using either kind of membrane alone.

However, two practical separation schemes can be proposed by using the two "imperfect" membranes together in cascade designs. In the first scheme Cascade 1 in Figure 1 is used. Stage 1 is installed with the SR membranes and stage 2 with the CA membranes. Most of CO_2 is removed in the permeate stream of the first stage. The reject stream from this stage will then be fed to the second stage with the permeate from the second stage being the product. The second scheme uses Cascade 2. The first stage uses CA membranes to remove most of N_2 and the second stage uses SR membranes to separate CH_4 from CO_2.

The strategy for designing the first stage is quite different from that for the separation of the most permeable species. Assume for the following discussion that the objective is to recover 64% of CH_4 from the feed mixture using Cascade 1. An obvious design constraint is that the reject stream from the first stage should contain more than 64% of CH_4 present in the feed. The second stage can then recover the CH_4 in the reject from the first stage to satisfy both the overall recovery and purity requirements. Since CH_4 has an intermediate permeability toward the SR membrane, very little change in its concentration takes place in the first stage. Consequently, neither the amount of recovery in the first stage nor the membrane area of this stage significantly affects the CH_4 concentration in the reject stream (the feed to the second stage). This fact is shown in Figure 7 by the family of constant-recovery curves (80, 70 and 67%) located in the narrow concentration range of 0.4 to 0.5. The CO_2 concentration in the reject stream, however, is affected by the CH_4 recovery to a greater extent, as indicated by the corresponding curves on the left side of Figure 7.

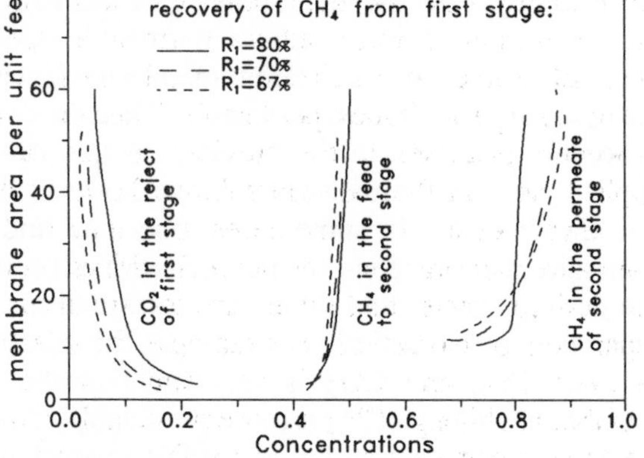

Figure 7. Cascade Using Different Membranes

Since CH_4 and CO_2 have similar permeabilities towards the CA membrane housed in the second stage and both species have much larger permeabilities than N_2, the separation taking place in that stage will primarily be between these two components and N_2. In order to meet the purity requirement, the CO_2 composition in the reject stream from the first stage should be less than 0.2, considering the fact that CO_2 will be inevitably concentrated along with CH_4 in the second stage. For example, when the membrane area of the first stage design is 10 m^2 /(mol/s), the reject stream from the first stage contains less than 15% of CO_2. While the area of the first stage will certainly have some effects on the design of the second stage, the recovery of the first stage is the predominant factor. Therefore, we choose to use an area of 10 m^2 /(mol/s) in the first stage for the following discussion of effects of first-stage recovery on the cascade design.

When the first stage is designed to recover 80% CH_4, the overall area needed for the second stage to enrich CH_4 to a concentration lower than certain level (about 0.8 in this example) is smaller than what would be needed if the recovery of the first stage is less than 80% (70 or 67%, for example). If the final CH_4 purity is higher than 0.8, the total membrane area requirement will be the least if the first stage recovers only 70% instead of 80% of CH_4. This reversal in membrane area requirement is basically caused by the following opposite effects. First, the required CH_4 recovery of the second stage decreases as the first-stage recovery increases. Lower recoveries require smaller membrane areas. Second, the CO_2 concentration in the feed to the second stage increases as the CH_4 recovery in the first stage increases. The separation of CH_4 from N_2 thus becomes less efficient due to the presence of more CO_2 which has a similar permeability as CH_4 in the CA membrane. The amount of CO_2 determines the practical upper CH_4 enrichment limit. Increasing the first-stage CH_4 recovery lowers the practical enrichment limit of the second stage. The competition between the two opposite effects is reflected by the crossing of the curves on the right-hand side of Figure 7, which represent the overall area requirement.

It is clear from the above discussion that the key parameter in this design is the first stage recovery. The choice of this parameter should be based on the final product purity requirement. It is not clear, however, how to determine an optimal recovery for the first stage so that the total membrane area requirement will be a minimum. The rules of thumb used for the separation of the most permeable species do not apply for the separation of the component with an intermediate permeability.

CONCLUSIONS

The separation of multicomponent gas mixture by recycle permeator cascades has been investigated using the diffusion model. The separation efficiency depends not only on the membrane selectivity and feed to permeate pressure ratio, but also on the design configuration. When the most permeable species is the desired product, high permeate (or reject) concentration requirements can be more efficiently satisfied by using cascade configurations instead of a single-stage design. The near optimal design of a two-stage cascade for recovering the most permeable species can be achieved when the two stages are designed for the same recovery and separation factor. The same design principle should be applicable to the recovery of the least permeable species.

When the product species is not most permeable in one membrane and has a very small permselectivity in another, a practical two-stage cascade can still be designed to accomplish the separation task. The recovery of CH_4 from CO_2-CH_4-N_2 mixture using both CA and SR membranes is used to demonstrate how a practical separation scheme can be designed with the two "imperfect" membranes. The important design factor in this case is the recovery of the intermediate species from the first stage.

NOTATION

- A membrane area
- f recycle fraction
- P_r feed/permeate pressure ratio
- R recovery
- x_f feed composition of the desired species
- y permeate composition of the desired species
- α_1 permselectivity of species 1

LITERATURE CITED

1. Gardner, R. J., R. A. Crane and J. F. Hannan, Chem. Eng. Prog., 73(10), 76(1977).
2. Weber, W. F. and W. Bowman, Chem. Eng. Prog., 82(11), 23(1986).
3. McReynolds, K. B., Chem. Eng. Prog., 81(6), 27(1985).
4. Kneieriem, M. Jr., Hydrocarbon Processing, 59(7), 65(1980).
5. Bollinger, W. A., D. L. MacLean and R. S. Narayan, Chem. Eng. Prog., 76(10), 27(1982).
6. Stookey, D. L., C. J. Patton and G. L. Malcolm, Chem. Eng. Prog., 82(11), 36(1986).
7. Shelden, R. A. and E. V. Thorpson, J. Membr. Sci., 19, 39(1984).
8. Thorman, J. M. and S. T. Hwang, Chem. Eng. Sci., 19, 15(1978).
9. Rangarajan, R., M. A. Mazid, T. Matsuura and S. Sourirajan, Ind. Eng. Chem., Process Des. Dev., 23, 79(1984).
10. Chern, R. T., W. J. Koros and P. S. Fedkiw, Ind. Eng. Chem., Process Des. Dev., 24, 1015(1985).
11. Stern, S. A., J. E. Perrin and E. J. Naimon, J. Membr. Sci., 20, 25(1984)
12. Chen, S., Y. K. Kao and S. T. Hwang, J. Membr. Sci., 26, 143(1986).
13. Kao, Y. K., S. Chen and S. T. Hwang, J. Membr. Sci., 32, 139(1987).
14. Boucif, N., A. Sengupta and K. K. Sirkar, Ind. Eng. Chem., Fundam., 25, 217(1986).
15. Hwang, S. T. and S. Ghalchi, J. Membr. Sci., 11, 187(1982).
16. Sengupta, A. and K. K. Sirkar, AIChE J., 33(4), 529(1987).
17. Sengupta, A. and K. K. Sirkar, J. Membr. Sci., 39, 61(1988).
18. Ghalchi, S. "Methane separation by a continuous membrane column", MS Thesis, University of Iowa, (1980).
19. Kothe, K. D., S. Chen, Y. K. Kao and S. T. Hwang, J. Membr. Sci., in press (1989).
20. Schell, W. J., Hydrocarbon Processing, Aug. 43(1983).

INTEGRATION OF MEMBRANE AND PSA SYSTEMS FOR THE PURIFICATION OF HYDROGEN AND PRODUCTION OF OXO ALCOHOL SYNGAS

K.J. Doshi, R.G. Werner and M.J. Mitariten ■ UOP Engineering Products, Old Sawmill River Road, Tarrytown, NY 10591

This paper presents the proper integration of a PSA and membrane system for the production of Oxo-alcohol synthesis gas. Operational aspects and economic comparison with a traditional PSA system are presented. The integrated system results in lower cost and increased reliability.

This paper presents an improved method for the ratio adjustment of hydrogen/carbon monoxide synthesis gas used in the manufacturing of oxo-alcohols. Assumed is a two step, staged process. In the first stage of production, an aldehyde is produced from a gas mixture of approximately one part hydrogen and one part carbon monoxide. In stage two, hydrogenation of the aldehyde with pure hydrogen results in the direct formation of the desired oxo-alcohol product. The overall hydrogen-to-carbon monoxide ratio required is two-to-one although this varies slightly depending upon the specific oxo-alcohol product being produced.

The complete oxo-alcohol plant has several main sections: (1) The synthesis gas section produces a raw stream of hydrogen and carbon monoxide which is sent to, (2) The purification and ratio adjustment section, followed by (3) The final synthesis reactor section that incorporates the downstream facilities to convert the hydrogen and carbon monoxide into the final oxo-alcohol products. These sections are further described below.

(1) Raw hydrogen and carbon monoxide synthesis gas is usually made by either steam reforming or partial oxidation of hydrocarbons. Steam reforming of natural gas gives a product stream with an excess amount of hydrogen in an approximately three-to-one H2:CO ratio. Partial oxidation of natural gas produces a product stream with an H2:CO ratio of approximately two-to-one. Partial oxidation of heavier hydrocarbons shifts the reaction toward larger percentages of carbon monoxide. The H2:CO ratios achieved by either process vary somewhat depending upon the hydrocarbon content of the raw feed gas and other process conditions. Either system requires ratio adjustment and further clean-up before the proper quality gas can be sent downstream to the synthesis reactors.

(2) Steam reforming and partial oxidation both produce streams with substantial carbon dioxide content. Carbon dioxide is undesirable in the downstream reactors and hence must be removed before further syngas processing. Carbon dioxide is most often removed by liquid absorption (either aqueous ethanolamine or potassium carbonate scrubbing). The carbon dioxide removed represents a carbon loss from the system (carbon which comes from the feedstock being reacted to make the hydrogen and carbon monoxide). Depending on the value of the feedstock, the carbon dioxide can be either disposed of at this stage, or recycled back to the raw feed for further recovery. Recycle of the carbon dioxide shifts the steam reforming or partial oxidation reaction equilibrium toward the production of larger percentages of carbon monoxide and lesser percentages of hydrogen.

(3) Since the oxo-alcohol process discussed here is a staged system, requiring a H2:CO ratio of one-to-one in the first stage and pure hydrogen in the second stage, either synthesis system will also require further ratio adjustment to meet the needs of oxo-alcohol production. Methods for this adjustment are discussed below:

RATIO ADJUSTMENT VIA IMPORT OF CARBON DIOXIDE

One method of achieving the required 1:1 ratio of hydrogen to carbon monoxide for aldehyde production is to import carbon dioxide into the partial oxidation unit (in a similar manner oxygen can be imported into a second stage of steam reforming). The import of the external stream allows production of a higher percentage of carbon monoxide and a lower percentage of hydrogen. The amount of imported gas can be adjusted so that a ratio of 1:1 of hydrogen-to-carbon monoxide is achieved (refer to Figure 1).

SYNTHESIS RATIO ADJUSTMENT VIA PRESSURE SWING ADSORPTION (PSA)

A common and much more flexible method of adjusting a hydrogen to carbon monoxide ratio after CO2 wash is through the use of a pressure swing adsorption (PSA) system (refer to Figure 2). The PSA system operates by adsorbing carbon monoxide at high pressure. A high purity hydrogen product is produced at high pressure. Carbon monoxide and unrecovered hydrogen are removed from the PSA unit at reduced pressure and recompressed to aldehyde synthesis pressure. Since the PSA unit normally operates at a higher hydrogen recovery rate than needed to simply reject hydrogen, a portion of the feed stream is bypassed around the PSA (at feed pressure) and combined with the compressed PSA tail gas. The hydrogen produced by the PSA unit contains only a few parts per million of carbon monoxide and is used in the downstream hydrogenation of the aldehyde.

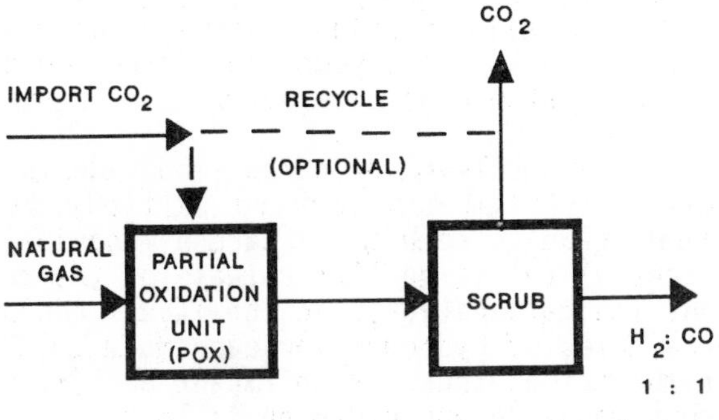

Figure 1. Ratio Adjustment Via Import of CO2

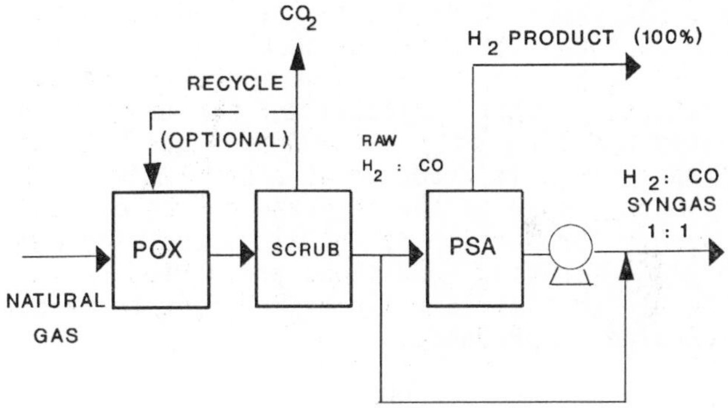

Figure 2. Ratio Adjustment Via PSA

Importing an external stream may have drawbacks. The cost of the external stream must be included in the overall project economics and may increase the unit production cost of the plant. Operating the steam reformer or partial oxidation unit in this manner also requires the availability of a high purity hydrogen stream for hydrogenation (in step 2) of the aldehyde. The importation of the two external streams has a direct impact on the economics of oxo-alcohol production. The dependence upon the supply of external gas streams not under the control of the oxo-alcohol plant is a factor more difficult to quantify.

When treating partial oxidation off-gas, the PSA recovers all the carbon monoxide in the feed and directly rejects the proper amount of hydrogen. There is no loss of either synthesis gas component and all the hydrogen is recovered at high purity.

A steam reformer feed provides hydrogen in quantities greater than the stoichiometric requirements for the production and hydrogenation of the aldehyde. The PSA operates in a manner similar to that of the partial oxidation case above, and the excess hydrogen produced is available for export to other hydrogen consumers.

The PSA system in a single step makes the high purity hydrogen product at high pressure as well as the proper synthesis gas ratio. There is no loss of carbon monoxide from the system and the resulting overall recovery of both the hydrogen and carbon monoxide is 100%.

An economic drawback of the PSA system used alone is the relatively expensive tail gas compression needed. PSA tail gas is produced at low pressure and uses a large compressor to boost the tail gas to the required aldehyde synthesis pressure.

SYNTHESIS RATIO ADJUSTMENT WITH A MEMBRANE SYSTEM

Another method to achieve the desired hydrogen-to-carbon monoxide ratio adjustment is the use of a membrane system (refer to Figure 3). The mixed stream from the partial oxidation unit or steam reformer enters the membrane unit after CO2 wash at high pressure, and the excess hydrogen in the feed gas permeates through the membrane wall to a pressure lower than the feed. The permeate hydrogen requires further purification for use in the hydrogenation step. Synthesis gas with the proper ratio is produced at approximately the same pressure as the feed and can be directly used for production of the aldehyde. The membrane operating conditions are varied so that the stoichiometric ratio of hydrogen-to-carbon monoxide is produced.

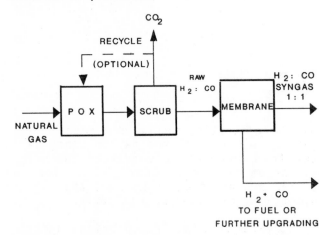

Figure 3. Ratio Adjustment With A Membrane System

The membrane system used alone is a simple and low cost unit for the ratio adjustment of synthesis gas, however its low purity reject hydrogen stream may be a significant economic drawback for the system for two reasons. The enriched hydrogen stream contains carbon monoxide, which varies between two and ten percent in concentration, dependent upon the membrane performance characteristics and process conditions. Carbon monoxide is a catalyst poison in the hydrogenation of the aldehyde and use of the hydrogen permeate requires removal or conversion of the carbon monoxide. The rejected carbon monoxide is usually lost from the system.

The quantity of oxo-alcohol product is dependent upon the amount of carbon monoxide in the feed. Thus for a fixed production rate of oxo-alcohol, a membrane system requires a larger raw feed gas flow to compensate for the losses of carbon monoxide into the permeate hydrogen stream. The membrane capital cost is small in proportion to that of the steam reformer or partial oxidation system, but the carbon monoxide loss increases the equipment size of the latter and the hydrocarbon feed rate needed to make a given product rate. The larger partial oxidation or steam reformer unit and larger raw gas production requirements generally eliminate any savings due to the lower cost of the membrane system compared to a PSA system plus compressor.

The high level of carbon monoxide in the hydrogen stream can require multiple stage methanation to remove the carbon monoxide and to control the large heat release of the reaction. The methanation reaction itself consumes three moles of hydrogen for each mole of carbon monoxide; thus a significant part of both carbon monoxide and hydrogen is lost.

Due to the difficulties of this methanation step a stand alone membrane system may alternatively import an external hydrogen stream, further increasing the cost and complexity of the system.

SYNTHESIS RATIO ADJUSTMENT WITH AN INTEGRATED MEMBRANE PLUS PSA SYSTEM

As described above the major drawback of the membrane system when used alone is the penalty associated with the loss of carbon monoxide with the hydrogen permeate. The major drawback of the PSA system when used alone is the power and cost penalty associated with the compression of the PSA tail gas. Integration of the two units into a single process system for oxo-alcohol production gives a unique way to avoid or minimize these drawbacks. The close combination of the two unit operations reduces the negative features of either one used alone, and gives a ratio adjustment system with strengths and capabilities which are enhanced significantly over those of the sum of its parts.

The integrated system is represented in Figure 4. Raw feed exits the steam reformer or partial oxidation unit and enters an absorption unit where the carbon dioxide is removed. The carbon dioxide may be recycled as feed to the reactor or rejected from the system for other uses.

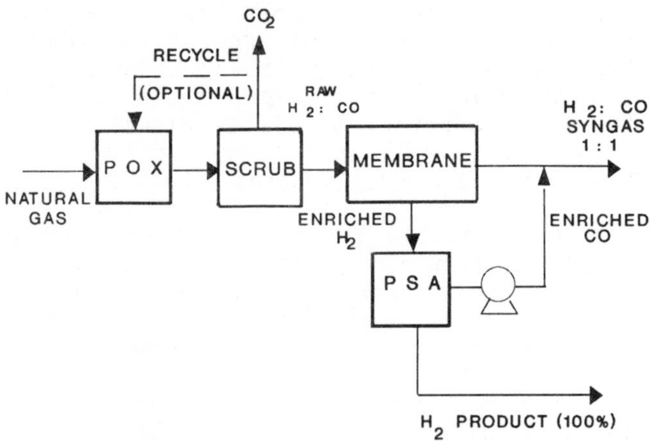

Figure 4. Integrated Membrane Plus PSA System

The carbon dioxide free stream consisting of mostly hydrogen and carbon monoxide enters the membrane unit where a non-permeate stream enriched in carbon monoxide is produced at essentially feed pressure. The hydrogen is removed as the reduced pressure permeate stream and fed to a PSA unit where the carbon monoxide is adsorbed and a high purity hydrogen stream is produced for hydrogenation of the aldehyde.

The tail gas of the PSA unit contains carbon monoxide as well as any unrecovered hydrogen. This stream is compressed to aldehyde synthesis pressure and combined with the high pressure non-permeate stream from the membrane unit. Ratio adjustment is achieved simply by varying the amount of high purity hydrogen extracted from the overall system. The unique benefits of the integrated system are described below:

In the integrated system all the available feed is introduced into a membrane unit for the production of an enriched hydrogen permeate stream. This step does not significantly differ from that of a stand alone membrane unit. The non-permeate comprises the major portion of the synthesis gas and is available at essentially feed pressure; thus the predominant portion of the oxo-alcohol synthesis gas does not require further compression.

Membrane area and cost increase as hydrogen recovery increases. The membrane unit here operates at a relatively low hydrogen recovery rate and is of compact design and low cost. The non-permeate produced from the membrane is enriched in carbon monoxide but the ratio of hydrogen-to-carbon monoxide is not yet set at that required for aldehyde production.

The permeate from the membrane system is enriched in hydrogen to a level typically greater that 90 %. Compared to a PSA system treating the raw hydrogen and carbon monoxide, this stream is more attractive for upgrading in a PSA unit due to the lower level of carbon monoxide to be adsorbed and the smaller flow rate than that of the raw synthesis gas. The stream is easily upgraded in the PSA unit so that the PSA is of a relatively smaller size and cost. Since the PSA unit produces hydrogen at essentially feed pressure, it is easy to meet the pressure required by the hydrogenation step, and no further hydrogen purification is required.

A Significant benefit is the reduced quantity of PSA tail gas produced by the integrated system. The large quantity of tail gas produced was the major drawback of the stand-alone PSA unit. A PSA unit treating the membrane permeate stream generates a tail gas less than half the size. The resulting saving in compressor size and power is a major advantage of the integrated system.

The tail gas from the PSA is compressed and recombined with the membrane non-permeate stream. Operation in this manner results in similar performance to that of a stand-alone PSA unit, i.e., 100% recovery of both carbon monoxide and hydrogen.

The integrated system results in a lower overall capital cost compared to a PSA system used to supply both the hydrogen product and the synthesis gas. The major capital advantage is a substantially smaller compressor which allows for both a reduction in capital and operating costs. A comparison of a stand-alone PSA and an integrated membrane plus PSA is presented in Table 1.

TABLE 1. Comparative Economics
PSA Alone versus Membrane + PSA
Integrated System

Basis

Raw Feed Rate, MM SCFD	20.0
Synthesis Gas Rate, MM SCFD	14.4
Hydrogen Product Rate, MM SCFD	5.6
Raw H_2/CO Gas Pressure, psig	420
Aldehyde Reaction Pressure, psig	400
Hydrogen Use Pressure, psig	150

Compositions, Mol-%	Raw H_2/CO	H_2 Prod.	Syn. Gas
H_2	63.4	99.999	49.1
CO	35.4	10 ppm	49.1
Ar/N_2	0.4	trace	0.6
CH_4	0.8	trace	1.2
CO_2	trace	trace	trace
H_2O	Sat'd	Dry	Sat'd

	PSA Only	Membrane + PSA
Compression Required, BHP	1080	415
Separation Equipment Cost, MM USD	1.425	1.375
Installation Cost, MM USD	0.175	0.225
Installed Compressor Cost	0.864	0.332
Compressor Operating Cost, MM USD (3 years-8000 hr/yr-5 cents/kW)	0.966	0.371
Total Capital Cost + 3 yrs Operation, (MM USD)	3.430	2.303

The membrane system operation is based upon a high pressure feed stream with hydrogen produced at lower pressure. The pressure profile of a PSA system is the opposite of a membrane unit in that hydrogen is produced at high pressure and impurities are removed at lower pressure. The inherent reaction pressure level of the two stage oxo-alcohol process is another reason for the synergy of the integrated membrane and PSA. The ability to consume the synthesis gas at high pressure and, more importantly, the utilization of hydrogen at lower pressure makes the integration particularly attractive.

SYSTEM OPERATION

In the production of oxo-alcohols the control of the H2:CO ratio is critical. Very precise control is achieved through a flexible system similar to that used in the PSA-only systems.

The flow rates of hydrogen and carbon monoxide in the feed stream are measured and recorded. From these flow rates the amount of hydrogen that must be rejected from the system is calculated, and used to set the hydrogen product rate from the PSA unit.

The design of the ratio control system for the integrated membrane/PSA system takes into account a variety of process considerations such as: the raw feed gas composition, membrane aging, selected H2:CO ratio required for the production of different oxo-alcohol products, as well as turn-up and turn-down conditions, and process upset situations.

The flexibility of the PSA system combined with knowledge of the proper integration requirements of each process reduces the sensitivity for maintaining a constant amount of carbon monoxide in the membrane permeate.

The system is designed to operate automatically with minimal operator attention.

A particulate feature of the integrated system is its ability to offer enhanced onstream factors without large cost penalty. In the event of a shutdown of any of the three processing units (membrane, PSA or compressor), the overall system can continue to make syngas, albeit at some reduced efficiency.

A failure of the compressor triggers an adjustment of the operating conditions so that the membrane unit directly produces the proper ratio of hydrogen-to-carbon monoxide as synthesis gas. The PSA's operation is also adjusted so that the tail gas is routed to fuel and the PSA continues to upgrade the permeate to high purity hydrogen. The overall system's operation under these upset conditions allows for operation approaching design capacity, albeit at lower efficiency.

Operation of the synthesis gas system during a PSA shutdown is more difficult than the case above. During an unscheduled outage on the PSA unit, it is possible to operate the membrane to produce the proper syngas ratio and to continue to make the aldehyde intermediate. Storage facilities for the aldehyde will allow continued operation of the plant.

An external source of acceptable purity hydrogen (if available) would allow for oxo-alcohol production.

A membrane shutdown triggers the control system to send the raw hydrogen plus carbon monoxide gas into the PSA directly. The system is automatically adjusted to increase the feed pressure to the PSA to that of the raw feed. Increasing this pressure does not offer any benefits of higher pressure hydrogen but it does increase the capacity of the PSA unit. The PSA system produces a hydrogen product and ratio adjusted synthesis gas at about half the capacity of normal operation.

CONCLUSIONS

The integrated synthesis gas ratio adjustment system described in this paper offers lower overall capital costs compared to the PSA only case while at the same time reducing the operating expenses of oxo-alcohol production. In addition to the lower costs, the integrated system offers the potential for higher on-stream factors, and greater operating flexibility with less installed-spare equipment in the process train. As a final and important benefit, the integrated system is a simple and straightforward separation scheme that directly produces the required product gases at the proper compositions without feedstock losses, i.e., 100% recovery of hydrogen and carbon monoxide.

The principles outlined in this paper apply to the upgrading of other synthesis streams. A complete knowledge of the membrane and PSA units allows the proper economic decision of a stand-alone system versus an integrated system as well as the design for a practical integrated system where appropriate. UOP will start up a synthesis gas ratio adjustment plant, based upon advanced hollow fiber composite membranes and POLYBED PSA technology, similar to that outlined in this paper, in the fourth quarter of 1989.

Much attention has been paid to the integration of membrane units with other separation processes. In most cases the integrated process does not offer benefits that outweigh the added costs or complications. The integrated membrane plus PSA system for oxo-alcohol synthesis gas ratio adjustment is one case where the synergy of the two systems results in benefits not otherwise available.

THE SEPARATION OF HYDROCARBONS FROM WASTE VAPOR STREAMS

R.-D. Behling, K. Ohlrogge, and K.-V. Peinemann ■ GKSS-Forschungszentrum Geesthacht GmbH, Postfach 11 60, D-2054 Geesthacht, Federal Republic of Germany
E. Kyburz ■ Aluminium Rheinfelden GmbH, Postfach 11 40, D-7888 Rheinfelden (Baden), Federal Republic of Germany

Hydrocarbon vapors generated from industrial processes dispersed into air are contributing factors for the creation of photochemical smog. The separation of hydrocarbon vapor by means of membranes is in case of some applications a technically simple and economic process. A membrane vapor separation process with a following treatment of the retentate by catalytic incineration is introduced in this paper.

INTRODUCTION

The dispersion of organic vapor emissions into air are governed by new Clean Air Regulations. The first of March 1986 was the effective date for the implimentation of TA-Luft in the FRG. This sets new limits for all kinds of air pollutants and defines a schedule for the improvement of existing plants in order to meet the new source performance standards. Parts of this Clean Air Act are regulatory issues for general organic compounds and those volatile hydrocarbons classed as carcinogenic.

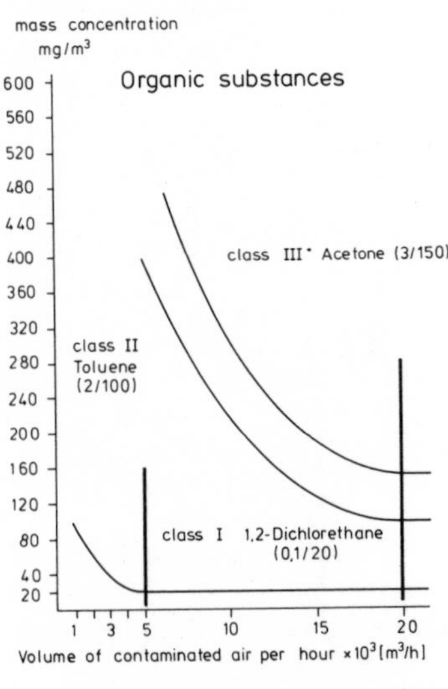

Figure 1: Organic substances

In Figures 1 and 2 the emission mass flow limits are shown in conjunction with the classification of the organic substance [1].

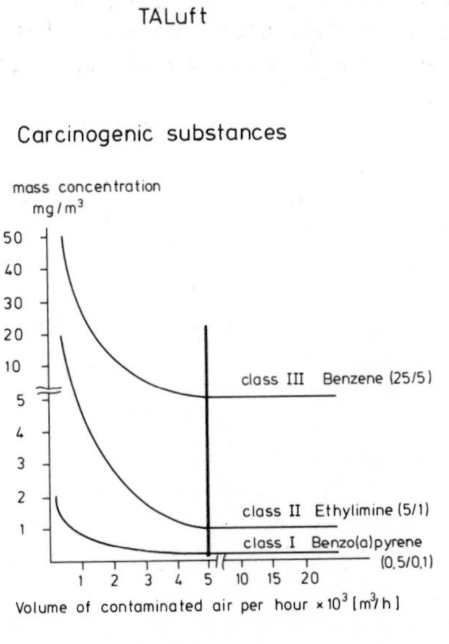

Figure 2: Carcinogenic substances

MEMBRANE FOR HYDROCARBON RECOVERY

The membranes which have been developed at GKSS Research Center for the recovery of hydrocarbon vapors are thin film composite membranes. A schematic diagram of such a composite membrane is shown in Figure 3.

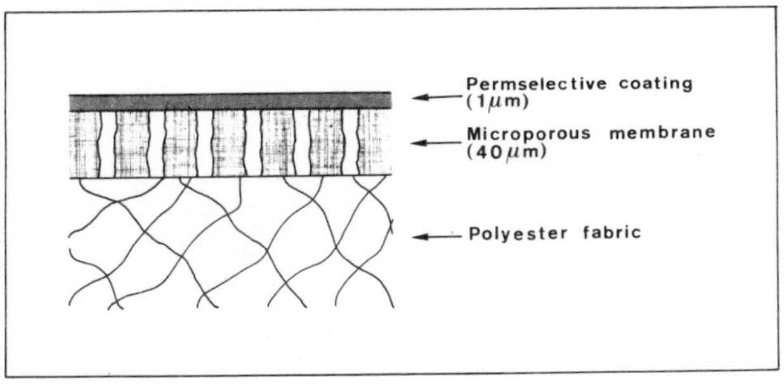

Figure 3: Schematic diagram of composite membrane

The membrane consists of three layers of different materials. The first layer is composed of an industrially produced, very porous, nonwoven polyester. The second layer is a microporous membrane which is applied on the first by the Loeb-Sourirajan technique. The microporous layer of the GKSS vapor recovery membrane consists of polyetherimide (Ultem, General Electric). Polyetherimide has been chosen instead of the commonly used polysulfone for two reasons:

1. The polyetherimide membrane has a higher permeability than a polysulfone membrane with a comparable pore size.
2. Polyetherimide is significantly more stable against fuel components like benzene and Toluene than polysulfone.

The third layer is the selective barrier layer. The thickness of the vapor recovery membrane ranges between 0.5 and 2 μm. The barrier layer is pore-free and consists of an elastomeric polymer. If a proper choice of the elastomeric material has been made, the membrane will have a preferential permeability for large organic molecules. This behaviour can easily be explained by the solution-diffusion model. In this model, it is assumed that gas at the high pressure side of the membrane dissolves in the membrane material and diffuses down a concentration gradient to the low pressure side of the membrane, where it is desorbed. The rate limiting step is the diffusion of the gas molecule through the membrane, which can be described by Fick's law. This leads to the equation

$$J = \frac{D \cdot k \cdot \Delta p}{l} \qquad (1)$$

In this equation J is the gas flux through the membrane, D is the diffusion coefficient of the gas in membrane, k is the Henry's law sorption coefficient, Δp is the pressure difference across the membrane and l is the membrane thickness.

Equation (1) implies that crucial parameters for the gas transport through a pore-free polymer film are diffusion- and solubility coefficent of the gas in the polymer. The diffusion coefficient of a molecule generally decreases with increase in size of the molecule, the solubility, however, commonly increases with the molecule size. Glassy polymers with stiff polymer backbones act like a molecular sieve. Small gases like hydrogen and helium can diffuse much faster through the rigid polymer backbone than gases with a larger diameter like hydrocarbons. An elastomer on the other hand acts more like a liquid. The transport of a gas through an elastomer is determined more by its solubility than by its diffusion coefficient. The high solubility of organic vapors in some elastomers is the reason for their high permeability.

This behaviour can be depicted as in Figure 4. The upper curve shows the permeability of the polyetherimide/silicone rubber composite membrane for oxygen, nitrogen and some hydrocarbons. The lower curve shows the opposite permeability pattern of the glassy polyetherimide membrane.

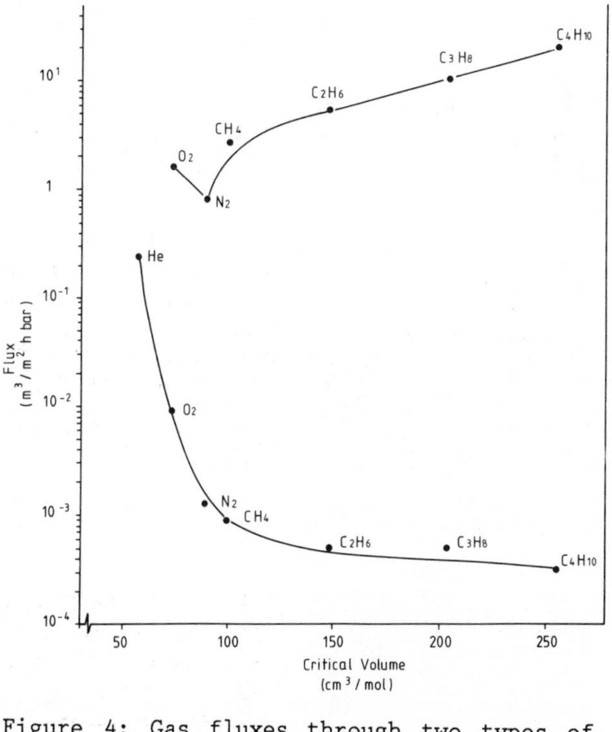

Figure 4: Gas fluxes through two types of polyetherimide/silicone composite membranes

The diffusion- and solubilty coefficient of oxygen, nitrogen and some hydrocarbons is illustrated in Figure 5. Figure 5 reveals that the diffusion coefficient of pentane in silicone rubber is 3.6 times smaller than the diffusion coefficient of oxygen, the solubility of pentane, however, ist 200 times higher than the solubility of oxygen in silicone rubber. The decrease of the pentane diffusion coefficient is overcompensated by the increase in its solubility. Therefore pentane permeates much faster through the membrane than oxygen.

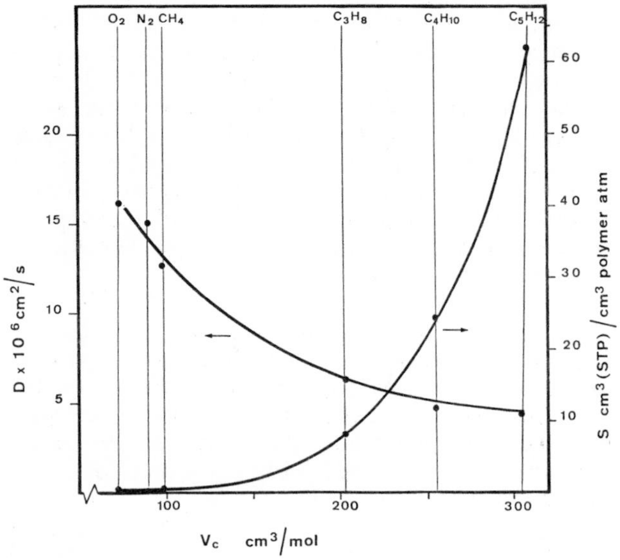

Figure 5: Diffusion- und solubility coefficients of various gases in silicone rubber vs. their critical volume, original data from [2, 3, 4]

Two important parameters of a membrane based vapor recovery unit are the temperature and the pressure of the feed gas. Whereas the permeation of inert gases through glassy polymers always increases with temperature, the opposite isoften true for the vapor permeation through elastomers. The temperature dependence of the solubility of gases in a polymer is best described by

$$S = S_o \exp(-\Delta H/RT) \qquad (2)$$

where S_o is a constant and ΔH is the heat of solution. The heat of solution becomes negative for the less volatile gases. This means that the solubility and consequently the permeability of these gases increase with decreasing temperature. This behaviour is shown in Figure 6 for the flux of butane, propane, ethane, oxygen and nitrogen through a polyetherimide/silicone rubber composite membrane.

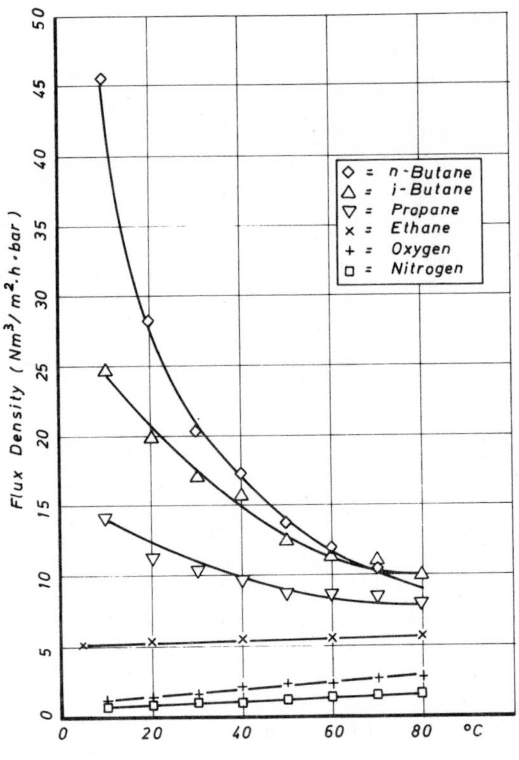

Figure 6: Flux density vs. temperature (P = 1230 mbar)

The second important transport parameter is the pressure of the feed gas. The flux density of nitrogen through a silicone rubber/polyetherimide composite membrane is practically pressure independent. The pressure dependence of the organic vapor flux, however, can be quite dramatic.

Figures 7 and 8 show the pressure dependence of hydrocarbons and various organic vapors through a silicone rubber/polyether-

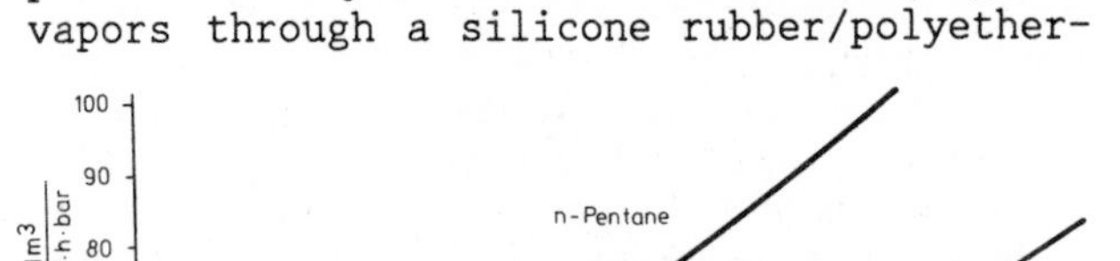

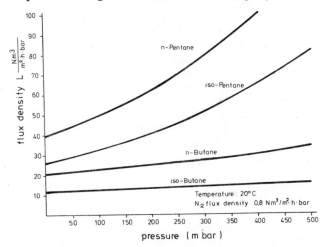

Figure 7: Butane and pentane flux density vs. pressure

imide membrane. Higher sorptions at high organic vapor pressures plasticize the membrane and increase the solvent's diffusion coefficient.

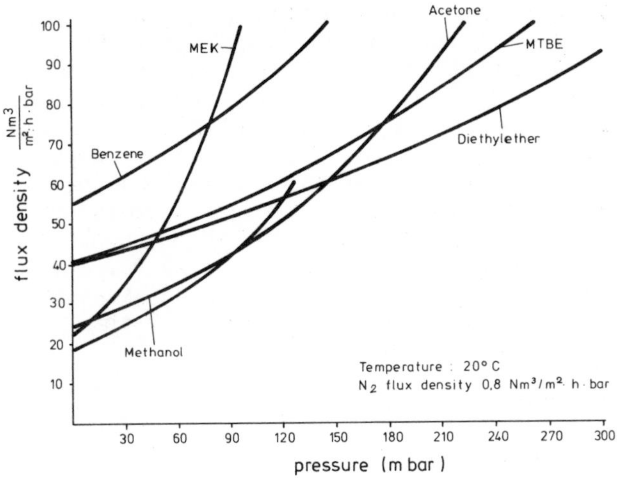

Figure 8: Fluxes of various organic vapours vs. pressure

Membrane module configuration

The flat sheet membrane module GS5 (Figure 9) is designed for applications with low pressure drop at the permeate side. Another advantage of this system is the provision to adjust the height of the feed channel to the requirements of feed viscosity and feed concentration [5].

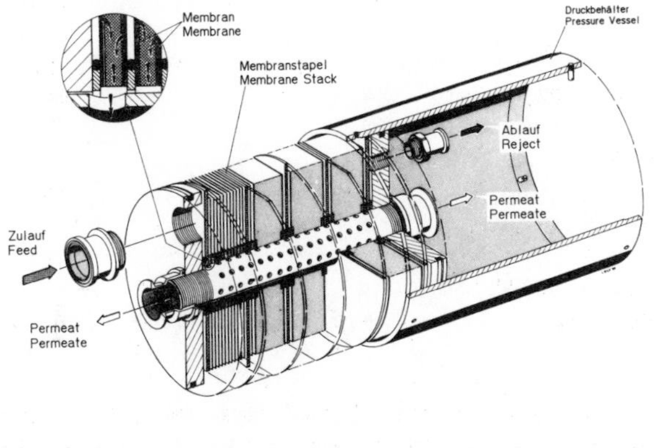

Figure 9: Flat sheet membrane module GS5

The GS-module consists of a tube like pressure vessel and a stack of membrane assemblies comprising of numerous membrane envelopes. The membrane envelopes are constructed using two round flat sheet membranes sealed at the cutting edges. Woven and non-woven materials are placed between the membranes to form the free volume for an unrestrained permeate drainage. The membrane envelopes are positioned along the central permeate tube and held in place by feed spacers between the envelopes. The permeate tube is used as a tension rod and has a thread at both ends where the final flanges and adapter sleeves are mounted. Gaskets between two membrane envelopes seal the feed side against the permeate side. The initial stress for the sealing is provided by the adjustment of the final flanges.

The feed flow is introduced into the module via the front flange, turns round by means of a baffle plate and flows over each membrane in the stack. The flow to the membrane can occur in parallel, in sequence or as a combination of the two depending on the layout requirements. The depleted stream leaves the module at the end of the membrane stack. The permeate which penetrates into the membrane envelope flows to the central permeate tube.

With regard to permeate pressure and permeate volume the permeate tube can be designed for one sided or both sided permeate drainage. This is to reduce the pressure drop as a result of friction at high flow velocities.

Process description

Organic compounds in tank off gases from fuel depots vary in concentration in the range of 100 g/m³ to 1500 g/m³. The composition of the organic vapor depends on fuel specification, e.g. regular or premium gasoline, summer or winter quality.

The main components are:
Butane / Butene
Pentane / Pentene
and minor components are:
Methane, Ethane, Propane, Hexane, Heptane, Toluene, Benzene.

In lead free grades, in particular methanol, ethanol, TBA, MTBE and TAME are used as octane number enhancers and may in addition be found as low level component. The layout of the demonstration plant (Figure 10 [6]) is based on an avarage hydrocarbon concentration of 37 vol% (approx. 1000 g/m³) and a feed gas flow of 300 m³/h.

Volatile hydrocarbons emitted from tank trucks during the filling procedure are fed into a gasometer (1). The gasometer is used as a buffer to homogenize the gas volume by alternating tank loading operations. A rotating piston compressor (2) pumps the tank off gases to the first membrane stage.

For safety reasons the concentration of hydrocarbons of the compressed gas stream must be above the upper explosion limit. The hydrocarbon concentration of the feedstream is measured by an infrared analyser. If the HC-concentration falls below a defined limit gaseous hydrocarbons are injected to enrich the feedstream.

The feed pressure of the first membrane stage is 2 bar, the vacuum at the permeate side 200 mbar. With regard to the calculations based on membrane data, the membrane area is determined to be 80 m^2. The stage cut comes to 0.72, the pressure ratio is 10. The retentate has a flow volume of approx. 80 m^3/h and a hydrocarbon concentration of 0.42 %.

A rotary vane vacuum pump (4) pumps the permeate to a screw compressor (5). The permeate is compressed to 10 bar to liquefy the hydrocarbons. The gas stream leaving the condenser (6) has a residual concentration of hydrocarbons depending on the final pressure and temperature. This stream is introduced in a second membrane stage (7) with 15 m^2 membrane area.

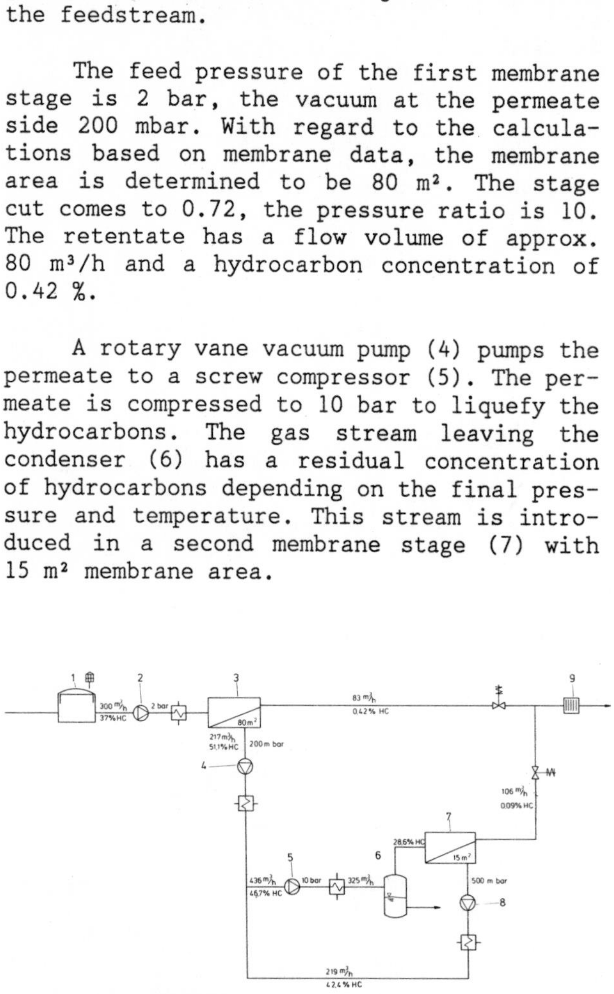

Figure 10: Demonstration plant

Because of the high feed pressure a pressure ratio of 20 can be achieved with a vacuum of 500 mbar at the permeate side (8). The hydrocarbon concentration in the retentate stream is calculated to be 0.09 vol%. The hydrocarbon enriched permeate stream is mixed with the permeate stream from stage 1 and fed back to the suction pipe of the screw compressor of the pressure condensation unit. Both retentate streams are combined resulting in a total flow of approx. 190 m^3/h with an average hydrocarbon concentration of 0.23 vol%. To meet the required emission standard of 150 mg/m^3 a post treatment of retentate (9) is necessary.

Tank farms have particular requirements. Very often electric power is the only energy supply. Accordingly a catalytic incinerator is then chosen to oxidize the hydrocarbons [7]. The operation temperature to start the catalytic inceneration is 200 °C. For safety reasons the surface temperature of the incinerator is limited to approx. 250 °C. Both temperatures depend on the hydrocarbon content of the gas stream. By a rule of thumb one gram of hydrocarbons causes a temperature increase of 50 °C. Therefore, the layout of the membrane stage must be tailored to the requirements of the catalytic incinerator.

Economics

a) Layout data

feed flow	= 300 m^3/h
average HC content	= 37 vol%
average HC density	= 2.6 kg/m^3
recovery rate	= 99.5 %
total recovery HC	= 290 kg/h
caloric value of recovered HC	=44.000 KJ/kg

The theoretical energy content of the recovered hydrocarbon amounts to 3500 kW. At a electric power conversion efficiency of 0.4 approx. 1400 kW could be produced.

b) Units of power of the main components

	engine power [kW]	shaft power [kW]
roots compressor	24	14.2
screw compressor	90	79
vacuum pump 1	37	30
vacuum pump 2	18.5	17.5
auxiliary pump	5	3.5
total installed capacity	174.5	total power consumption 144.2

An additional 20 kW are installed for the electrical heat exchanger to start the catalytic incinerator. This calculation shows that the plant could be operated with approx. 10 % of the converted energy from the recovered hydrocarbons.

c) Estimation of the value of recovered hydrocarbon

The VTG tank farm has a calculated

transshipment of 500.000 m³/a. This is the annual volume which has to be processed in the recovery plant. With regard to the average HC content of 37 % and the estimated recovery rate of 99.5 % the annual amount of recovered hydrocarbon is 185.000 m³ or 480.000 kg. This represents a value of 384.000 DM at a price of 0.8 DM/kg HC.

The experience gained from the operation of this demonstration plant will provide data for the improvement of the process modelling and for the design of recovery units for other applications. The engineering of this plant was tailored to an almost unmanned fuel depot without the infrastructure and skills of an integrated refinery.

Membrane separation can also operate in combination with an adsorption process or an combustion engine to produce the energy for the plant. A concentration of 80-50 g HC/m³ (approx. 3-2 vol%) enables the operation of an combustion engine [8]. This gas engine can be coupled to a generator to supply electrical power to operate the condensation unit. The flue gas of the combustion engine will meet the clean air standards.

A dual system of membrane separation and activated carbon adsorption or molecular sieves can provide advantages in plant performance and operation reliability. Molecular sieves are sensitive to water vapor and at high levels of organic load of some components in the waste stream there is a risk of spontaneous inflammability of the activated charcoal because of the high adsorbing heat. The adsorption properties of the adsorbent can be influenced by some adsorbed hydrocarbons with a high boiling point. These substances often show poor desorption behaviour. The membrane stage can be used to homogenize loading peaks of the feed stream or to separate components of complex mixtures.

References

[1] Allgemeine Verwaltungsvorschrift zum Bundesimmissionsschutzgesetz (Technische Anleitung zur Reinhaltung der Luft - TA Luft) vom 27. Februar 1986.

[2] Robb, W.L.: Thin Silicone Membranes - Their Permeation Properties and some Applications. Ann. N.Y. Acid. Sci., 146 (1960) 119.

[3] Barrie, J.A.; Mundy, K.: Gas Transport in Heterogeneous Polymer Blends. J. Membrane Sci., 13 (1983) 175-195.

[4] Barrer, R.M.; Barrie, J.A.; Raman, N.K.: Solution and Diffusion in Silicone Rubber. Polyer, 3 (1962) 595-614.

[5] Hilgendorff, W.; Kahn, G.; Kaschemekat, J.: Vorrichtung mit Membranen. (Patent application, DE 3507908 A1).

[6] Behling, R.-D.; Hattenbach, K.; Ohlrogge, K.; Peinemann, K.-V.; Wind, J.: Verfahren zum Austrag organischer Verbindungen aus Luft/Permeatgasgemischen. (Patent application, P 38 06 107.4).

[7] Schwefer, H.J.: Verfahren und Einrichtung zur Aufbereitung eines Kohlenwasserstoff-Luftgemisches. (Patent application, EP 269 572 AZ).

[8] Frey, G.: Verfahren zum Entfernen von Verunreinigungen aus einem Gas. (Patent application, EP 0222 158 AZ).

A MEMBRANE SYSTEM FOR THE SEPARATION AND RECOVERY OF ORGANIC VAPORS FROM GAS STREAMS

J.G. Wijmans and V.D. Helm ■ Membrane Technology and Research, Inc., 1360 Willow Road, Suite 103, Menlo Park, CA 94025

A newly developed membrane process to treat airstreams contaminated with volatile organic compounds is described. The process uses multilayer composite membranes which are selective for organic vapors over nitrogen and oxygen. The membranes are assembled into spiral-wound modules specifically designed for this particular membrane application. A pilot plant containing 15 m^2 of membrane area has been constructed and is being operated at MTR's facilities. The membrane system removes organic vapors from air streams and produces a secondary stream enriched in the organic. The organic is recovered from this stream by condensation. The system is best suited for applications where the organic vapor content is too high to make carbon adsorption economically feasible, but too low to permit incineration. The membrane process is also capable of removing organic vapors from gas streams other than air, such as natural gas streams.

Contamination of effluent airstreams with dissolved organic solvents such as naphthas, ketones, toluenes, esters and chlorinated hydrocarbons is an important environmental problem. Strict environmental legislation now requires industry to clean up these airstreams to the extent that final plant discharges are essentially solvent-free. Unfortunately, volatile organic compounds are difficult and expensive to remove even when present at the level of a few hundred ppm. This paper describes a recently developed membrane process that allows substantial cost savings over alternative volatile organic compound removal techniques.

MEMBRANE VAPOR SEPARATION

Membrane vapor separation is a low-pressure membrane process for separating organic solvents from air. A semipermeable composite membrane performs the separation. This membrane is incorporated into membrane modules which allow a large membrane surface area to be packed in a small volume.

In a vapor separation system, a contaminated airstream is introduced into an array of membrane modules. Organic solvents are preferentially drawn through the membrane by a vacuum pump. The solvent is condensed and removed as a liquid. The purified air stream is removed as the residue. Transport through the membrane is induced by maintaining the vapor pressure on the permeate side of the membrane lower than the vapor pressure of the feed airstream. This vapor pressure difference is achieved by means of a rotary-vane vacuum pump specially designed to treat streams containing organic solvents.

Air and organic solvents permeate the membrane at a rate determined by their relative permeabilities and the applied vacuum. Typically, the permeate is 10-50 times more concentrated in organic solvent than the feed air. This allows the volatile organic solvent to be easily removed by cooling and condensing the liquid.

Vapor Separation

The heart of the membrane vapor separation technology is the type of membrane developed. The MTR membranes are composite structures made by coating a tough, relatively open microporous support membrane with a very thin, dense film. The support membrane provides mechanical strength and the thin, dense coating performs the separation. Because the dense coating is so thin, very high permeation rates are achieved. The feed air contacts the nonporous surface of the membrane so that surface fouling and pore plugging do not occur.

An example of the performance of a vapor separation membrane is shown in Figure 1. Depending on the vapor used, the permeate contains 15 to 100 times more organic than the organic in air feed mixture.

Membrane Modules

MTR incorporates its membranes into spiral-wound membrane modules for treating solvent-contaminated air. The spiral-wound module design is illustrated in Figure 2. In this module design, a membrane envelope is wound around a porous central collection pipe.

Mesh spacer material is used to form channels for the feed air and the permeate vapors. The feed air is circulated laterally through the module. As the airstream passes across the membrane surface, the organic solvent fraction preferentially passes through the membrane and enters the permeate channel. This permeate vapor spirals inward to the central permeate collection pipe and then passes to the permeate condenser.

The membrane modules can be connected in serial and parallel flow arrangements to meet the capacity and solvent removal requirements for nearly all airstreams.

VAPOR SEPARATION SYSTEM

A membrane vapor separation system includes membrane modules, feed air compressor, permeate vacuum pump, permeate condenser and condenser chillers. The permeate flow across the membrane is obtained by developing a partial vapor pressure difference between the feed air and the permeate side of the membrane. The permeate flow of solvent across the membrane is proportional to this vapor pressure difference. A difference in vapor pressure is easily generated by using a compressor to increase the pressure on the feed side of the membrane and a vacuum pump to reduce the pressure on the permeate side of the membrane. The actual operating conditions used depend mainly on the composition of the feed. The membrane process is extremely easy to operate and contains no moving parts other than the pumps.

The vapor separation system come in a number of designs for various applications. The simplest system is a single-stage unit shown schematically in Figure 3. In this unit, feed air is compressed to 1-2 atm pressure and passed through the membrane modules. The treated air is discharged to the atmosphere or recycled to the process. The permeate vapor, enriched in the organic solvent, is passed to a condenser. The condensed solvent is transferred to a solvent holding tank.

A single-stage system is generally able to remove 80-90% of the solvent from the feed air and produce a permeate that has five to ten times the concentration of the feed gas. This degree of separation is adequate for many applications. For example, we have recently been evaluating the use of vapor separation systems to concentrate the butane produced as an off-gas in production of polystyrene foam packaging materials. Only 80% recovery of the butane is required in this application to meet EPA requirements and the goal is to concentrate the butane to a sufficient level to allow it to be used as a supplemental fuel in an existing boiler. The design of this unit and the estimated capital and operating costs are shown in Figure 3.

In many applications, although 80-90% removal of the solvent from the feed air is adequate, further concentration of the solvent is required to make efficient condensation of solvent from the permeate possible. This is the case if the initial feedstream is relatively dilute. In these cases, a two-stage unit is used, and the permeate from the first stage becomes the feed to the second stage. This configuration allows solvent enrichments of 50-to 100-fold to be achieved. Because the feedstream to the second stage is very much smaller than the feed to the first, the second stage is normally only 10-20% as large as the first stage.

The typical separation that can be achieved with a two-stage system is illustrated in Figure 4. The system is designed to treat a 500-scfm stream containing 1% 1,1,1-trichloroethane. In this application, the cost of operating the unit is less than the value of the removed solvent, so that there is a small positive cash flow.

A final system design is used when more than 90% removal of solvent from the feedstream is required. A two-step process, in which the residue from the first step is subjected to further treatment is then employed. Solvent removals of 95-99% are easily achieved. However, the second step required to reduce the feed concentration from 10% to 1% of the initial value is as large as the first step required to reduce the feed concentration from 100% to 10% of the initial value. This type of system is, therefore, only used with high-value solvents, such as Freons.

An example of a two-step, two-stage process for recovery of a Freon from effluent air is shown in Figure 5. As shown in this example, 98% removal of the CFC-113 from the feed air is achieved. Although the cost of the process per scfm of feed treated is relatively high, the value of the Freon is sufficient to make this design economically attractive, achieving a payback time of only a few months.

COMPETITIVE PROCESSES

Membrane vapor separation is a newly developed technique for the removal and recovery of organic contamination from airstreams. Competitive processes already in the marketplace are carbon adsorption and incineration.

Incineration, because it involves burning the organic solvent vapors with a supplementary fuel, such as natural gas, is best suited to relatively concentrated streams which require less fuel. The process is simple, reliable and inexpensive to install. On the other hand, solvent recovery is not possible and non-flammable solvents such as halocarbons cannot be incinerated.

Carbon adsorption is a cyclic process in which three large beds of activated carbon are used. At any one time, two of the beds are used to remove solvent from the feed air while the third bed is being stripped with steam. Because of the cyclic nature of the process, relatively complex automatic valving is required, as well as a boiler for steam generation. The size of the carbon beds can be very large, especially with high-volume airstreams containing more than 0.2-0.5% solvent. The operating and capital costs of carbon adsorption plants increase with the solvent concentration of the feed gas. Carbon adsorption cannot be used successfully for all organic solvents. Some organic compounds, such as chlorinated solvents, are not stable during the steam regeneration cycle, leading to corrosion of the system. Other components, such as low boiling hydrocarbons, are not adsorbed by the carbon.

A comparison of membrane vapor separation with carbon adsorption shows that the membrane process is more economic if the organic compound concentration is relatively high and if the airstream to be treated is small. This is because the operating cost of carbon adsorption increases with increasing total organic compound feed flow rate, whereas the operating cost of the membrane process scales with the air feed flow rate. This makes the membrane process an eminent candidate to control organic compound emissions directly at the source point where dilution with air is minimized. This contradicts the old adage "the solution to pollution is dilution", under which volatile organic compound emissions are dealt with by moving enormous volumes of air in the range of 50,000 to 500,000 scfm. Recovering the organic compounds at the source will reduce the streams to be treated by several orders of magnitude.

SUMMARY

A newly developed membrane process is capable of removing and recovering valuable organic compounds from airstreams. The process is especially suited to treat low-volume, high-concentration streams and is more economic than carbon adsorption.

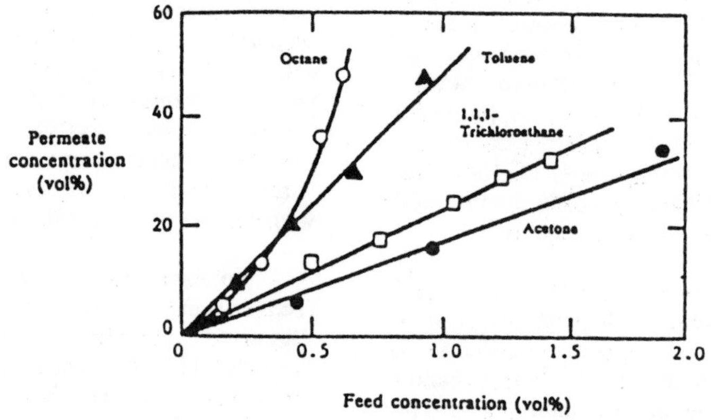

Figure 1. Concentration of organics in the permeate stream as a function of the organic concentration in the feedstream obtained with the MTR-100 vapor separation membrane.

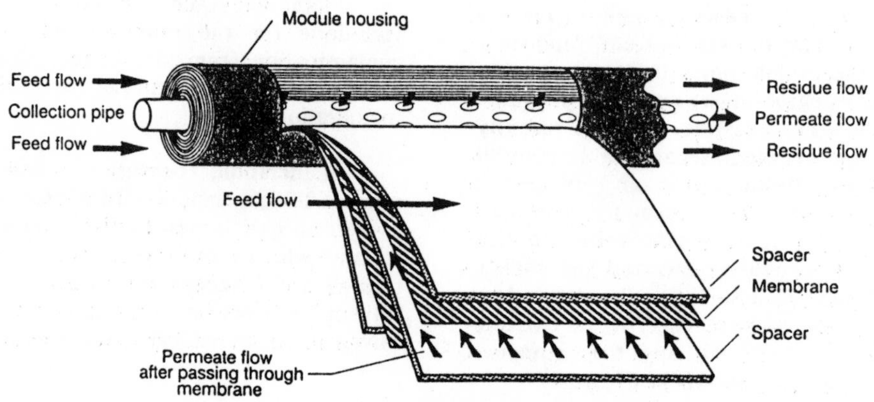

Figure 2. Spiral-wound membrane module.

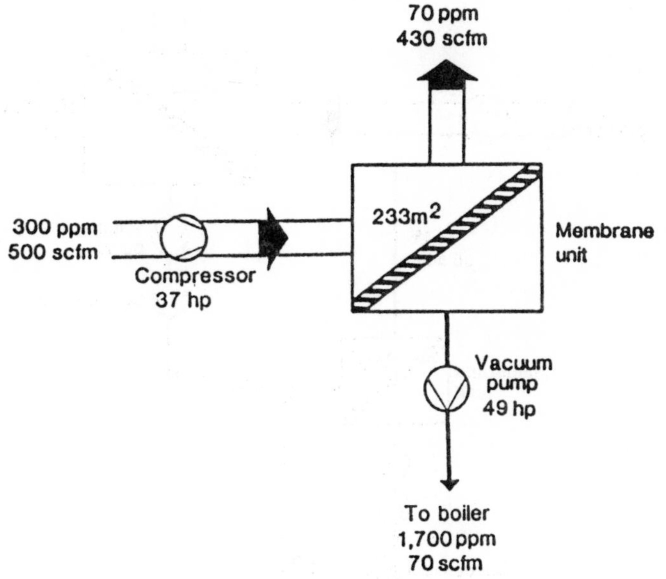

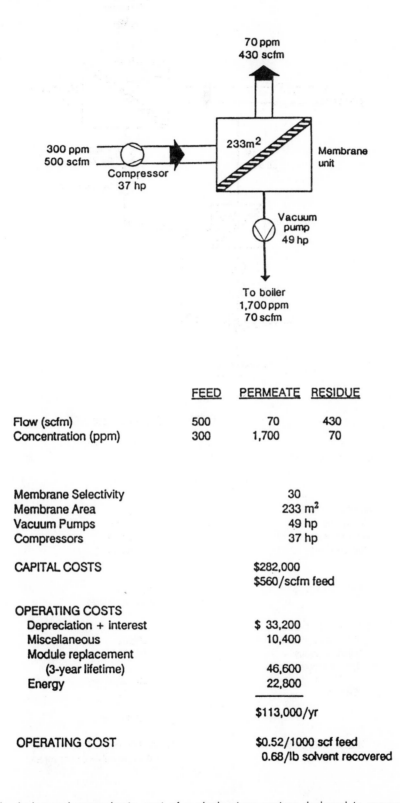

	FEED	PERMEATE	RESIDUE
Flow (scfm)	500	70	430
Concentration (ppm)	300	1,700	70

Membrane Selectivity	30
Membrane Area	233 m^2
Vacuum Pumps	49 hp
Compressors	37 hp
CAPITAL COSTS	$282,000
	$560/scfm feed
OPERATING COSTS	
Depreciation + interest	$ 33,200
Miscellaneous	10,400
Module replacement (3-year lifetime)	46,600
Energy	22,800
	$113,000/yr
OPERATING COST	$0.52/1000 scf feed
	0.68/lb solvent recovered

Figure 3. Schematic design and approximate cost of a single-stage system designed to remove 80% butane from an industrial off-gas for pollution control.

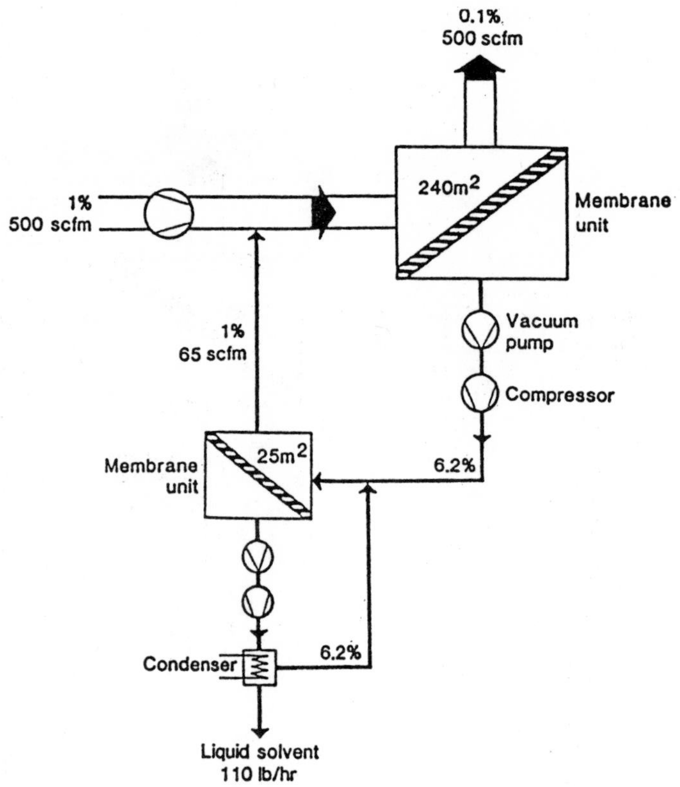

	FEED	PERMEATE	RESIDUE
Flow (scfm)	500	Liquid	500
Concentration (%)	1.0	–	0.1

Membrane Selectivity	30
Membrane Area	265 m^2
Vacuum Pumps	98 hp
Compressors	46 hp

CAPITAL COSTS	\$425,000
	\$850/scfm feed
OPERATING COSTS	
Depreciation + Interest	\$ 58,500
Miscellaneous	15,400
Module replacement (3-year lifetime)	53,000
Energy	38,100
	\$165,000/yr
OPERATING COST	\$0.76/1000 scf feed
	0.21/lb solvent recovered

Figure 4. Schematic design and approximate costs of a two-stage system designed to remove 90% 1,1,1-trichloroethane from degreasing bath off-gas.

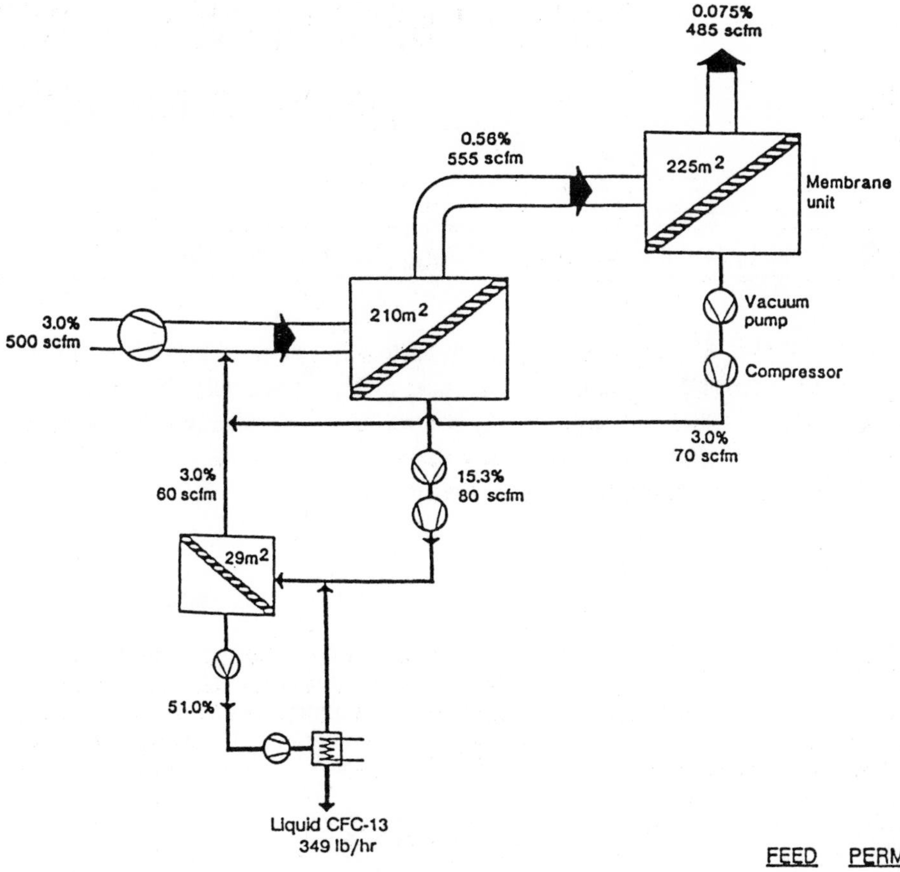

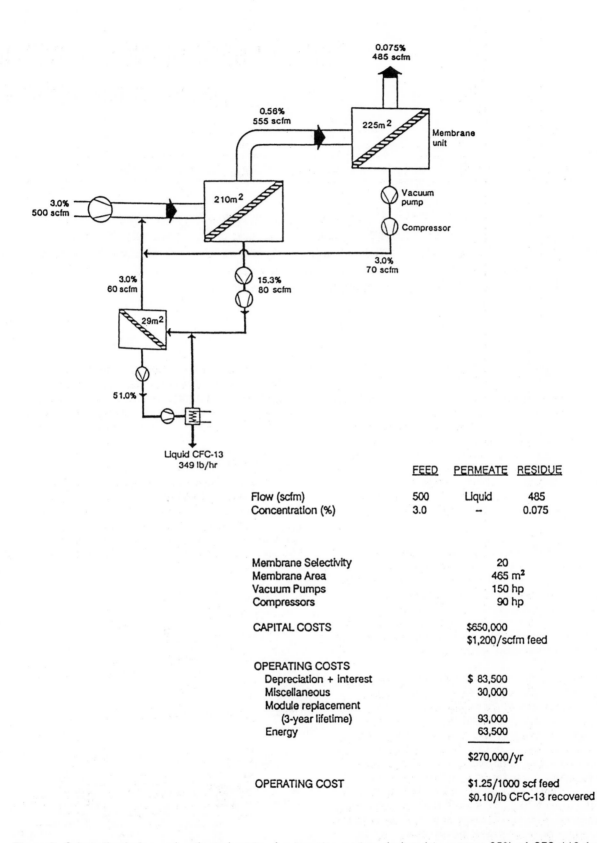

	FEED	PERMEATE	RESIDUE
Flow (scfm)	500	Liquid	485
Concentration (%)	3.0	–	0.075

Membrane Selectivity	20
Membrane Area	465 m^2
Vacuum Pumps	150 hp
Compressors	90 hp
CAPITAL COSTS	\$650,000 \$1,200/scfm feed
OPERATING COSTS	
Depreciation + Interest	\$ 83,500
Miscellaneous	30,000
Module replacement (3-year lifetime)	93,000
Energy	63,500
	\$270,000/yr
OPERATING COST	\$1.25/1000 scf feed \$0.10/lb CFC-13 recovered

Figure 5. Schematic design and estimated costs of a two-step system designed to recover 95% of CFC-113 from a 3% feedstream. The value of the recovered Freon far exceeds the operating cost per pound.

HIGH PERFORMANCE REACTOR USING A MEMBRANE

Y. Shindo, N. Itoh and K. Haraya ■ National Chemical Laboratory for Industry, Tsukuba, Ibaraki 305 Japan

A high performance membrane reactor, which is a double-tublar reactor equipped with a selective membrane tube as the inner tube, was proposed. Such a reactor makes it possible to obtain a product yield of a reversible reaction beyond its equilibrium value by continuous removal of the products during reaction. A palladium alloy tube was employed as the selective membrane, through which only hydrogen permeates. Experiments in the dehydrogenation of cyclohexane to benzene as a model reaction were carried out. It was shown that a marked increase in conversion over that at equilibrium could be achieved.

INTRODUCTION

It is well known that a reversible reaction can never be completed under ordinary conditions because the conversion attainable is limited by the thermodynamics. Let us consider the dehydrogenation of cyclohexane as an example of reversible gas-phase reaction. When this reaction reaches a state of chemical equilibrium at any fixed conditions, no further change in the composition with time can occur.

Michaels (1) pointed out that if the products (both or either) can be separated from the reacting mixtures by employing a selective gas separation membrane, the reaction will advance in the forward direction and finally will go to completion. Raymont (2) suggested a similar concept for increasing the decomposition yield of hydrogen sulfide, whose equilibrium conversion is quite low.

Such an idea has been tested by several investigators. Gryaznov (3) and Nagamoto et al. (4) studied dehydrogenation and/or hydrogenation using palladium and its alloy membranes. Carles et al. (5) attempted direct thermolysis of water vapor using a calcia-stabilized zirconia membrane. In these cases the membranes simultaneously function as catalysts. Kameyama et al. (6), Shinji et al. (7), and Itoh et al. (8) succeeded in selective separation of hydrogen from a reacting mixture using a microporous glass membrane. As a result, they obtained a higher conversion than that at equilibrium.

In this study, dehydrogenation of cyclohexane to benzene was investigated by means of a membrane reactor using palladium (9, 10). As a result, the dehydrogenation was promoted over its equilibrium conversion.

EXPERIMENTAL

Figure 1 shows a schematic diagram of the experimental apparatus. Details of a double-tube membrane reactor using a thin palladium alloy (palladium 77 %, silver 23 %) tube, 0.2 mm thick, 17.0 mm OD, and 140 mm long, is shown in Figure 2. Only hydrogen permeates

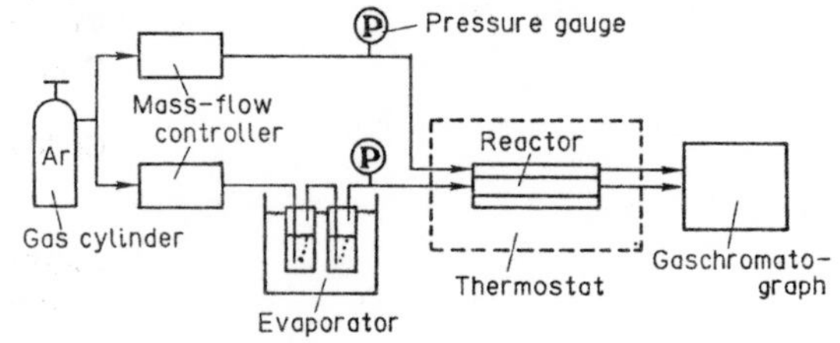

Figure 1. Experimental apparatus.

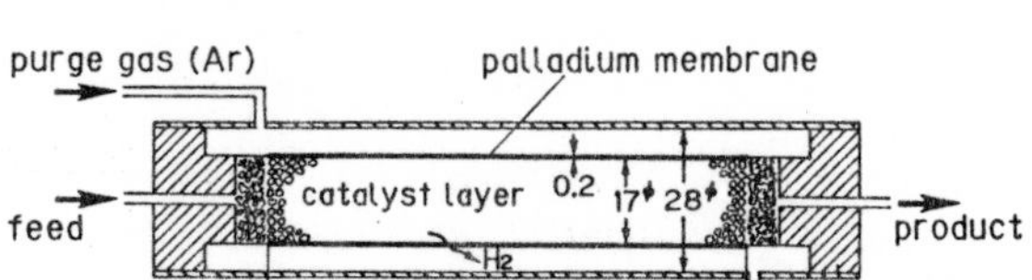

Figure 2. Details of membrane reactor.

through the palladium membrane. Platinum supported on cylindrical alumina (0.5 wt %, 3.3 mm OD, 3.6 mm high) were uniformly packed inside the membrane tube.

A 19.7 % saturated vapor of cyclohexane generated by flowing argon as a carrier gas through an evaporator was fed into the membrane reactor. Simultaneously argon was used as a purge gas to remove the hydrogen permeated to the shell side through the palladium membrane. The reaction temperature was 473 K, and the experiments were carried out at atmospheric pressure. The experimental conversion was determined by analyzing a concentration ratio of cyclohexane to benzene by means of a gas chromatgraph.

RESULTS AND DISCUSSION

Figure 3 shows experimental results. The solid curves show the calculated values. u_c^o is the flow rate of the cyclohexane at the inlet of the membrane reactor. v_a^o is the flow rate of purge gas argon. Conversion of 99.7 % was

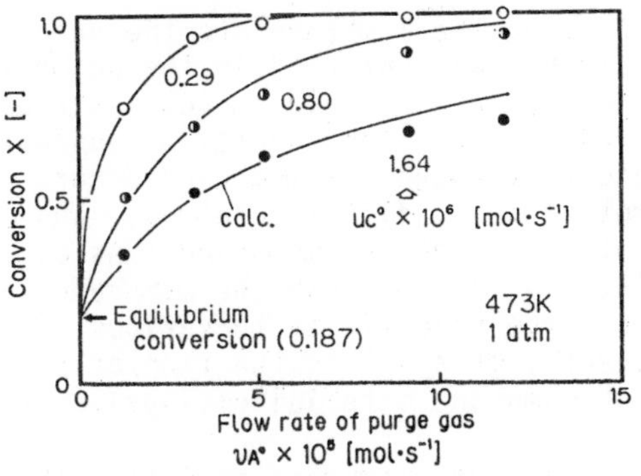

Figure 3. Experimental results.

attained; this value is much higher than at equilibrium, 18.7 %, which is the maximum conversion obtained by using an ordinary catalytic reactor. In this case, as only benzene leaves from the outlet of the reaction side, no additional separation apparatus between benzene and cyclohexane will be needed; this may lead to reduce opperating coasts. Instead of using a purge gas as described above, there is an alternative method for removing the permeated hydrogen. The pressure on the separation side can be maintained at a reduced level or vacuum. In such a case, very pure hydrogen can be obtained from the outlet of the separation side, so that direct hydrogen production in the course of the reaction would be possible.

CONCLUSION

A high performance reactor using a palladium membrane was proposed. Dehydrogenation of cyclohexane to benzene was taken as a model reaction. Nealy 100 % conversion was attained, where its equilibrium conversion was 18.7 %. This high performance membrane reactor can be used for any other dehydrogenation systems, which may include synthesis of styrene from ethylbenzene, shift-reaction of carbon monoxide and so on.

The principle of the high performance reactor using a membrane is the simultaneous operation of both reaction and separation. This principle can be applied to other decomposition systems, for instance, decomposition of water or carbon dioxide, when an oxygen separation membrane such as YSZ (yttria-stabilized zirconia) is used.

LITERATURE CITED

1. Michaels, A. S., Chem. Eng. Prog., 64, 31 (1968).
2. Raymont, M. E. D., Hydrocarbon Proc., 54, 139 (1975).
3. Gryaznov, V. M., Kinetics and Catalysis, 12, 640 (1970).
4. Nagamoto, H. and H. Inoue, Chem. Eng. Commun., 34, 315 (1985).
5. Carles, B. and J. F. Baumard, High Temperatures-High Pressure, 14, 681 (1982).
6. Kameyama, T., M. Dokiya, M. Fujishige, H. Yokokawa and K. Fukuda, Ind. Eng. Fund., 20 97 (1981).
7. Shinji, O., M. Misono and Y. Yoneda, Bull. Chem., Soc. Jpn., 55, 2760 (1982).
8. N. Itoh, Y. Shindo, K. Haraya and T. Hakuta, J. Chem. Eng. Jpn., 21, 399 (1988).
9. N. Itoh, AIChE J., 33, 1576 (1987).
10. N. Itoh, K. Miura, Y. Shindo, K. Haraya, K. Obata and K. Wakabayashi, Sekiyu Gakkaishi, 32, 47 (1989).

DEVELOPMENT OF MEMBRANE PERVAPORATION TRIM™ PROCESS FOR METHANOL RECOVERY FROM CH_3OH/MTBE/C4 MIXTURES

Michael S.K. Chen, Gregory S. Markiewicz, and
Kamesh G. Venugopal ■ Air Products and Chemicals, Inc., 7201 Hamilton Boulevard, Allentown, PA 18195-1501

This paper describes the successful technical development efforts using modified Separex cellulose acetate membranes for separating azeotropic liquid mixtures of methanol/MTBE/C4s. The patented TRIM™ process was conceived, laboratory-tested, process-evaluated, and market-tested in less than two years. A 4 gpm skid-mounted demonstration unit which has two membrane tubes each with two 2″ spiral wound membrane elements has been successfully operated at a MTBE plant site in the Gulf Coast. The TRIM process, Total Recovery Improvement for MTBE, represents the first major application of CA membrane technology to organic liquid separation markets.

MTBE, methyl tertiary-butyl ether, is a premium octane enhancer for motor fuel, growing worldwide as much as 20% per year due to lead phasedown and increasing premium gasoline demand. In the conventional MTBE production process, methanol in slight excess is used to achieve moderately high conversion of isobutene and to minimize side reactions. Unfortunately, methanol forms azeotropes with both MTBE and unreacted C4s, which are difficult to separate by distillation. The conventional process scheme is to take this azeotropic mixture to a debutanizer to produce a MTBE bottom product and a binary methanol/C4 azeotrope overhead. This stream is then subjected to a water wash to remove the excess methanol from C4s. Methanol/water mixture is distilled to recover the methanol for recycle. This conventional process is both capital and energy intensive.

In the TRIM process, a membrane pervaporation system operates by feeding a liquid mixture of methanol/MTBE/C4s to the membrane unit and selectively permeating methanol through the membrane as vapor. This is brought about by keeping the pressure on the permeate side far below the feed pressure either by a vacuum pump or by an ejector. The vapor permeate is recovered by condensation. The membrane unit can be located between the reactor and the debutanizer, or located on a sidedraw from the debutanizer column. The TRIM process is projected to realize attractive cost savings for both grassroots and retrofit applications, saving as much as 20% of the additional plant investment required for a high conversion MTBE plant.

Introduction

Air Products is commercializing a new process for the production of methyl tertiary-butyl ether, or MTBE, utilizing advanced membrane separation technology for organic liquids. MTBE has quickly become the favored blending component to improve octane in the gasoline pool, with demand growing by as much as 20% per year worldwide (1-3). Air Products' new process is known as the Total Recovery Improvement for MTBE, or TRIM™, process. The TRIM process represents Air Products' first adaptation of commercial Separex membranes to serve a growing market for liquid separations. The TRIM process is projected to offer attractive cost savings for both grassroots and retrofit MTBE plant applications.

Membrane Pervaporation

Pervaporation is a unit operation for separation of liquid mixtures based on selective membrane permeation (4-5). Pervaporation systems operate with liquid feed and residual streams, and a vapor permeate under vacuum (Figure 1). A liquid mixture is passed over a polymeric membrane, where one or more components pass through the membrane preferentially, evaporating at some point along the path (hence the name 'pervaporation').

Exiting the membrane are the vapor permeate stream enriched in the permeating components, and a liquid residual stream depleted in those components. Transport of components through the membrane occurs by absorption into, diffusion across, and desorption out of the membrane. As a result, the flux through the membrane, and the relative selectivity to various components, is a strong function of the type of membrane material employed.

For a given membrane material, the rate of transport of component i is controlled by the difference in component concentrations in the feed side and permeate side of the membrane via Fick's Law:

$$J_i = - D_i \frac{dc_i}{dx} \quad \text{(local)} \qquad \text{Equation 1}$$

$$J_i = - D_i \, (C_{i,feed} - C_{i,permeate}) \quad \text{(overall)} \qquad \text{Equation 2}$$

These concentrations are related to the equilibria at the feed-membrane and membrane-permeate surfaces. On the feed side, the concentrations in the membrane are assumed to be in equilibrium with the liquid mixture. On the permeate side, the concentrations in the membrane are in equilibrium with the component partial pressures in the vapor permeate. Because the equilibria are complex, it is convenient to express the overall driving force across the membrane in terms of vapor pressures, which can be determined for the feed side from equilibrium with the liquid mixture. Thus, the driving force can be expressed as:

$$(C_{i,feed} - C_{i,permeate}) \propto$$
$$(p_{i,feed} - yi\ P_{permeate}) \qquad \text{Equation 3}$$

To maximize driving force for a given feed mixture, the engineer can either raise feed temperature to increase $P_{i,feed}$, or lower the permeate pressure $P_{permeate}$. Since feed component vapor pressures are often relatively low, on the order of 50-100 torr, pervaporation systems are operated with a vacuum on the permeate to allow evaporation and thus increase driving force.

We've described the concentration gradient as proportional to the vapor pressures in Equation 3, because estimating the membrane component concentrations in any solution of these equations is very difficult. First, the diffusion coefficient Di is a function of composition across the membrane. Secondly, the equilibrium determinations of vapor pressures at the membrane surface must be made with the actual feed and permeate fluids. Such measurements are extremely difficult. For engineering purposes, then, we must often rely on empirical results from experiments to understand pervaporation membrane performance.

Pervaporation is different from distillation in that it does not depend upon component relative volatilities for separation. Thus, pervaporation is an attractive alternative to distillation for separating azeotropic mixtures, close-boiling compounds, and also finds application in removing small amounts of a component from a bulk liquid stream. On the other hand, because pervaporation requires heat input for evaporation and mechanical energy to maintain a vacuum, it is not competitive with distillation for bulk separation of liquids.

Conventional MTBE Process

MTBE is made by reacting methanol with isobutene from a mixed-C_4 stream, in the liquid phase over an acidic ion-exchange catalyst (Figure 2). The catalyst is so selective that only the isobutene in the C_4 feedstock reacts with the methanol. The reaction is equilibrium-limited. In commercial practice, reactor conversions of isobutene are typically 87-94% per pass, with molar ratio of methanol to isobutene slightly above 1.0 (6-8). The reactor effluent contains MTBE, mixed C4s, and unreacted methanol; methanol forms azeotropes with both MTBE and C4s. The reactor effluent is distilled to produce an MTBE bottoms product, and the methanol-C4 azeotrope overhead. The methanol is recovered from the C4s by water wash or mole sieve separation, and recycled. A flow diagram is shown in Figure 3.

Excess methanol can be fed to the reactor to increase isobutene conversion, with up to 5-6% additional conversion available. Excess methanol can also reduce side reactions, such as di-and tri-isobutene formation. Adding extra methanol, however, is a separation problem, because of the azeotropes involved. The extra methanol sent to the debutanizer column would end up in the MTBE product. When additional conversion is desired, a two-stage process is used with intermediate MTBE removal (Figure 4).

The isobutene feedstock comes from three sources: (1) FCC offgas, (2) steam cracker offgas, usually after butadiene extraction, and (3) isomerization and dehydrogenation of field butanes. The mixed C4 streams from these three sources vary both in isobutene content and composition of the other C4s (Table 1). Isobutene feedstock has a bearing on MTBE plant costs and on flexibility in handling methanol in the process. Higher isobutene content gives higher MTBE production per volume of feedstock, but reduces the amount of other C4s available to remove excess methanol from MTBE in the debutanizer.

TRIM Process for MTBE Production

Air Products' TRIM process can be employed to achieve high conversion of isobutene, while reducing the complexity of and investment required for the plant. The TRIM pervaporation system separates methanol from MTBE and C4s selectively, thus breaking the methanol azeotropes. In one version of the process, the TRIM pervaporation system is placed on the reactor effluent (Figure 5). The methanol to isobutene molar ratio is increased from 1.0 to 1.2, improving MTBE conversion about 5% per pass. The pervaporation unit removes methanol from the reactor effluent (reducing the concentration from about 5 wt% to 2 wt%) to the level normally fed to the debutanizer in the conventional process. The back end of the plant, debutanizer and methanol recovery, is unaffected by the increase in methanol to the reactor. Additional conversion to MTBE is obtained without the need for a second distillation step to handle the azeotropes. This TRIM scheme can replace the extra reactor-debutanizer stage in the conventional process, saving 10-15% or more of the additional capital for a new high conversion plant. For an existing standard conversion plant, adding a TRIM system can gain the additional 5% MTBE production, without affecting the existing distillation train. The additional MTBE production can pay back the investment required for the TRIM system in 18 months, or less in some cases.

An alternate configuration for the TRIM process takes a sidedraw from the debutanizer to the TRIM pervaporation system to remove methanol (Figure 6). This is possible because methanol accumulates in the debutanizer column, due to the VLE behavior of the system. A composition profile of the debutanizer shows two methanol concentration peaks, one above the feed tray and one below (Figure 7). By taking a sidedraw, removing methanol with the pervaporation unit, and returning the residual to the column, the same overhead and bottoms specifications can be met with more efficient column operation. In the limit, enough methanol could be removed from the column to significantly reduce the overhead methanol content, reducing methanol recovery costs or possibly even eliminating the methanol recovery section. Another benefit of the sidedraw: because the methanol concentration in the feed to the pervaporation unit is higher than in the first scheme, the driving force (and hence the flux) across the membrane is higher, reducing the required membrane area. If the methanol-isobutene ratio is increased to 1.2, the additional conversion is obtained, and the excess methanol can be removed and recycled in the sidedraw pervaporation unit with the advantages mentioned above. The sidedraw scheme could be particularly attractive for new plants saving as much as 20% of the additional capital investment required for the high conversion plant.

In some situations, the TRIM pervaporation system can be located on the debutanizer overhead or bottoms streams to remove methanol; these locations offer fewer advantages for improving efficiency or cost of the MTBE process.

The TRIM system could be applied to other ether production processes where alcohol azeotropes occur, such as tertiary-amyl ether (TAME) or ethyl tertiary-butyl ether (ETBE) processes.

TRIM Pervaporation System Development

The TRIM pervaporation system has been developed over the last two years, and Air Products has recently completed a field demonstration of the system. The TRIM membranes are an adaptation of our commercial Separex cellulose acetate gas separation membranes to liquid separation service. Our experience base with membranes allowed a rapid development of the TRIM technology, now entering commercialization.

The development program focused on three main tasks: determining membrane performance over the commercially important range of process variables for system design, optimizing the membrane element design, and conducting the field demonstration. Process variable ranges of importance commercially are shown in Table 2. Laboratory experiments were conducted to map flux and selectivity as a function of these parameters, and to provide a basis for comparison with field test data. In addition, membrane life tests are ongoing to determine any flux and selectivity changes over time.

The field demonstration test was conducted at a major MTBE producer's plant on the U.S. Gulf Coast. The test was run for five months, and was designed to evaluate the membrane system performance under actual operating conditions. A schematic of the membrane system for the field test is shown in Figure 8. Two parallel tubes were used that housed four commercial spiral wound pervaporation elements (Figure 9). The elements were 2" diameter by 40" inch length. A feed mix tank was used to allow us to vary the feed composition by mixing the various components in set proportions. The feed mixture was first pumped to the membrane tubes through a heater/cooler arrangement to control temperature. The residual mixture was returned to the mix tank. The permeate was removed from each end of the tubes so we could check individual performance of each element. The permeate was collected and compressed in a 10 HP vacuum pump, then condensed against cooling water and collected in the receiver tank. Any noncondensibles were vented and the condensed permeate pumped back to the feed mix tank. A separate tube was used to house two larger 4" diameter by 40" elements that were also tested as confirmation of performance in a larger configuration. Both the 2" and 4" elements were specially constructed to directly model the performance of a larger 8" element, which would be used in a commercial system.

Because the feed-to-permeate driving force for methanol transport is low, it is important to minimize backpressure in the permeate channel of the element to maximize overall driving force. Several proprietary design features have been incorporated in the Separex pervaporation element to accomplish this. Through these measures, the element internal permeate pressure drop is significantly reduced.

Membrane system performance is detailed in Figures 10-15. Relative methanol and total fluxes are plotted as a function of methanol driving force in Figure 10. As the performance model discussed earlier suggests, methanol flux increases with increasing methanol driving force (which is obtained by either increasing methanol concentration, raising temperature, or lowering vacuum pressure). Notice that total flux remains parallel with methanol flux, indicating that the flux of C4s/MTBE remains constant. As methanol driving force increases, then, selectivity for methanol also increases, and this is shown in Figure 11. Selectivity is defined as the ratio of methanol flux to C4 or MTBE flux. The selectivity over MTBE is extremely high, which is important to the overall process since little MTBE will be recycled to the reactor to impair equilibrium conversion. Methanol concentrations in the permeate are typically in the range 65-90 wt %, with 1-10 wt % methanol in the feed, with the balance being mostly C4s.

Comparison of demonstration skid methanol flux with laboratory data is shown in Figure 12. The laboratory data was obtained on a synthetic mixture using n-butane as the C4 component. Laboratory data were also obtained using a feed mixture from the MTBE plant, with C4s composition close to a steam cracker feed that contains a large amount of olefinic components. The methanol flux on steam cracker feed is higher than the synthetic feed mixture, as seen both in the laboratory test with plant feed as well as the demonstration skid. Higher methanol flux would be expected, because of the higher methanol vapor pressure in the steam cracker feed. This has also been observed with dehydro-based feed. Methanol selectivity to MTBE is very high and essentially the same for all feeds. Selectivity to C4s on steam cracker feed is lower (1 to 3) than with the synthetic mixture (7-15). C4 selectivity with dehydro feed is closer to the synthetic n-butane performance. These data point out the importance of testing with the actual feed to be separated.

The effect of feed temperature is shown in Figure 13. Here we have plotted data versus methanol content in the feed, to include the effect of temperature on driving force across the membrane. There is a practical limit for the maximum temperature of operation based on membrane stability. The effect of vacuum pressure at constant temperature is shown in Figure 14. This is also plotted against feed methanol percent to illustrate the effect on driving force. At these conditions, vacuum pressure has a marginal effect because the overall driving force changed by only 15%. Finally, the effect of different ratios of MTBE to mixed C4s is shown in Figure 15. Changing composition affects both component solubilities and diffusivities in the membrane and the effective methanol vapor pressure, increasing or decreasing methanol flux accordingly.

<u>Current Status of TRIM Process</u>

Air Products is currently finalizing system designs for the TRIM process, and is beginning commercialization. Patents have been granted or allowed in the United States on the TRIM process and on methods of preparation of the membrane, and are pending in several overseas countries.

Summary

Air Products' TRIM process offers attractive productivity improvements and cost savings for both existing plant operators and new plant owners. The TRIM pervaporation membrane system has been developed and field tested over the last two years, and is currently being commercialized. The TRIM system may also be applicable to the production of other ethers, such as tert-amyl ether (TAME), mixed ethers, and ethyl tert-butyl ether (ETBE).

Acknowledgements

The authors would like to express our sincere thanks to the management, engineering, and operations staff at our host MTBE plant, without whose support and patience we would have not been able to conduct the very successful demonstration test. The authors also thank Air Products for permission to publish our paper.

References

1. __________, "MTBE's Prospects Appear Very Strong," Chemical Marketing Reporter, 9 January 1989, pg. 3.

2. Debreczeni, E. "C_4 Olefins Market Feast of Famine?," 1988 DeWitt & Co. Petrochemical Review (DeWitt & Co., Houston, 23 March 1988), pp. XVI 1-16.

3. Ludlow, W. I., "MTBE - Continued Strong Growth," 1988 DeWitt & Co. Petrochemical Review, pp. VII 1-8.

4. Rautenbach, R., and Albrecht, R. "The Separation Potential of Pervaporation, Part 1; Discussion of Transport Equations and Comparison with Reverse Osmosis," Journal of Membrane Science, 25 (1985), pp. 1-23.

5. Neel, J. "Fundamentals and General Chemical Engineering Aspects of Pervaporation," Proceedings of First International Conference on Pervaporation Processes in the Chemical Industry, R. Bakish, Ed., (Bakish Materials Corp. Englewood, NJ, 1986), pp. 10-40.

6. Lee, A. K. K. and Al-Jarallah, Adnan, "MTBE Production Technologies and Economics," Chemical Economy & Engineering Review, 18, No. 9 (1986), pp. 25-34.

7. Friedlander, R. H., "Huls Methyltertiary butyl Ether Synthesis and Decomposition Processes," Handbook of Chemicals Production Process, R. A. Meyers, Ed., (McGraw-Hill, New York, 1986), pp. 1.13-1-5.

8. Beck, T. A., and Miller, D. J. "ARCO MTBE - The Maximum in Flexibility," 1988 DeWitt & Co. Petrochemical Review, pp. XIV 1-14.

9. Trotta, R. "The Snamprogetti MTBE Technology," 1988 DeWitt & Co. Petrochemical Review, pp. XV 1-15

Table 1.

Typical C4 Feedstock Composition for MTBE Plants

Composition (wt%)	FCC Offgas	Steam Cracker Offgas	Dehydrogenation of Butanes
C_3	1	<1	1
N-butane	15	9	4
Isobutane	34	2	41
Isobutene	13	45	45
1-butene	11	23	4
2-butenes	23	20	4
1,3-butadiene	1	<1	<1
C_5	2	-	-
	100	100	100

Table 2.

Tested Parameter Ranges of Commercial Interest for TRIM Pervaporation Membrane System

Feed Methanol Composition	0.5 - 10.0 wt%
C_4/MTBE Ratio	25:75 - 75:25 (depends on % isobutene in feed)
C_4 feedstock type	FCC, steam cracked, dehydro
Feed Temperature	85-135°F (30-58°C)
Permeate Pressure	5 -50 torr abs (68 - 680 kg/m^2)

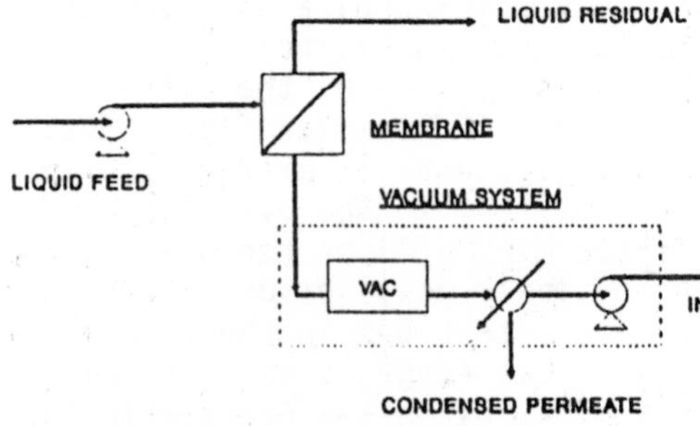

Figure 1. Membrane pervaporation.

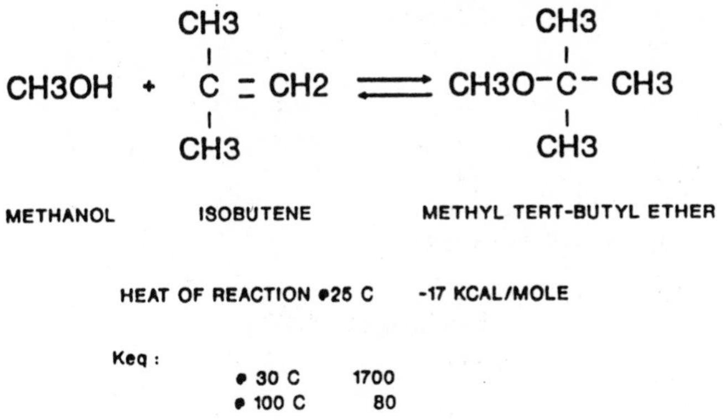

Figure 2. MTBE reaction chemistry.

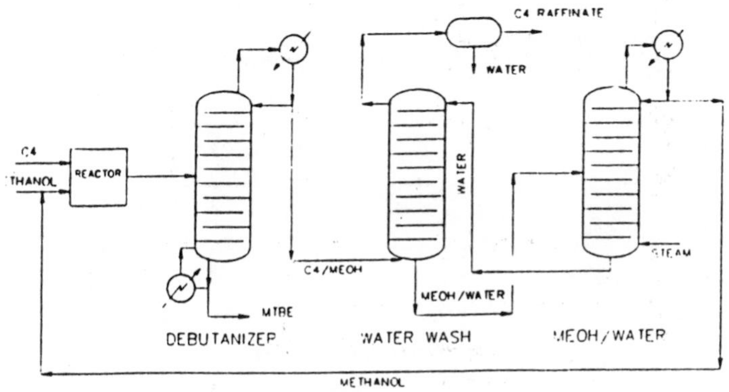

Figure 3. Standard MTBE process methanol recovery by water wash.

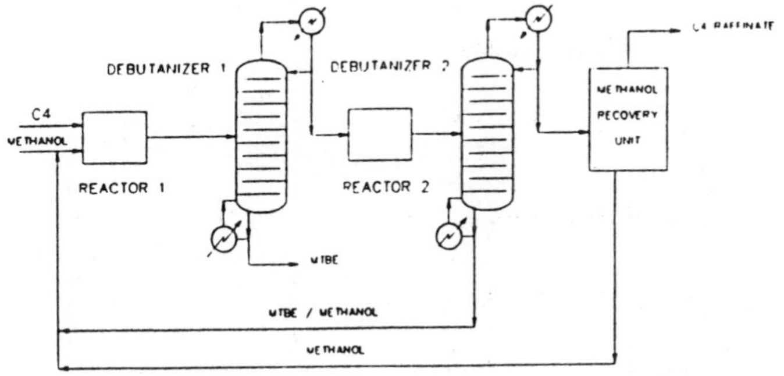

Figure 4. High conversion MTBE process.

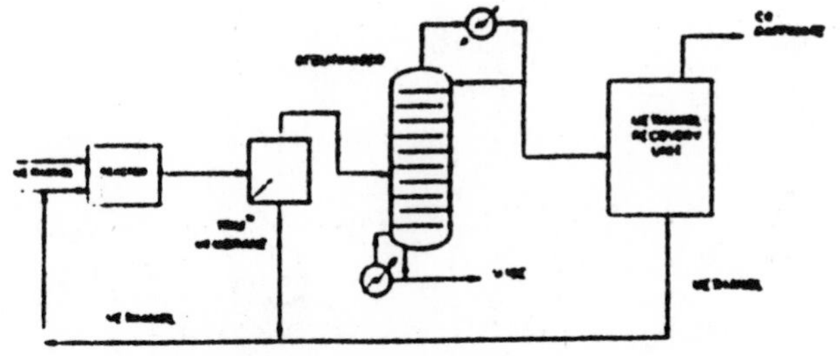

Figure 5. Trim process membrane before debutanizer.

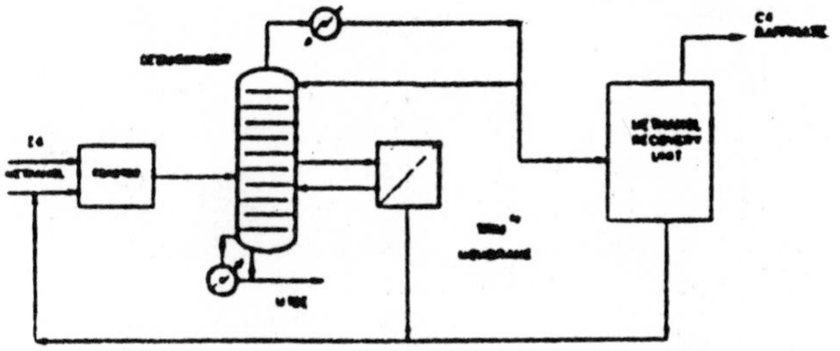

Figure 6. Trim process debutanizer side draw.

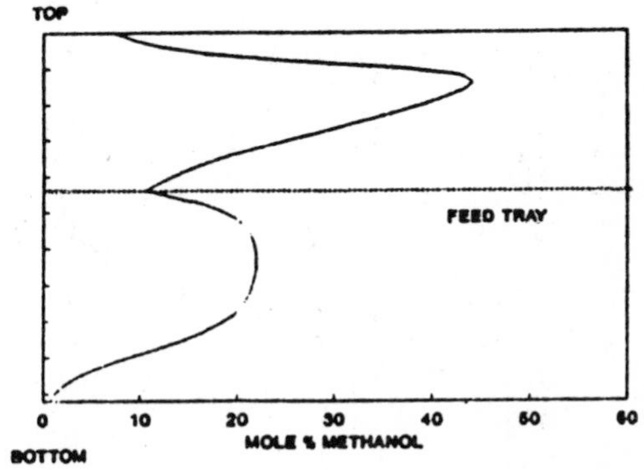

Figure 7. Debutanizer column profile liquid methanol % by tray.

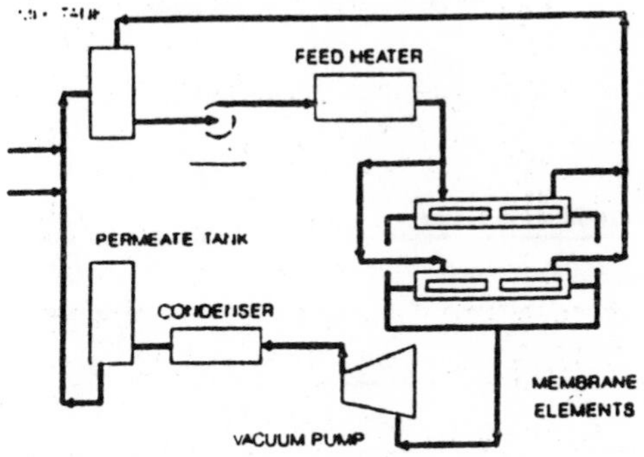

Figure 8. Trim membrane demonstration skid.

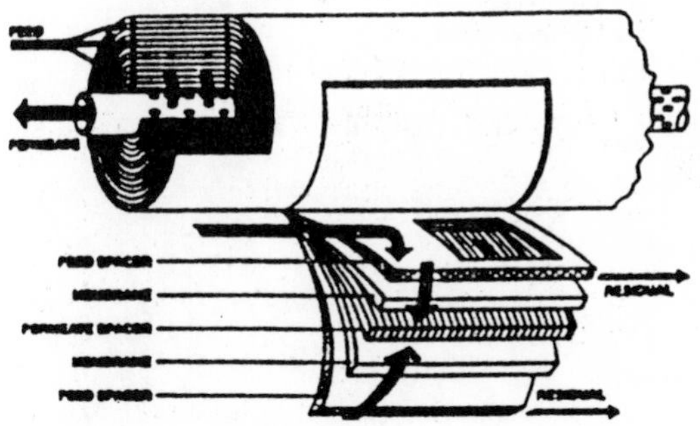

Figure 9. Spiral wound element configuration.

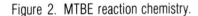

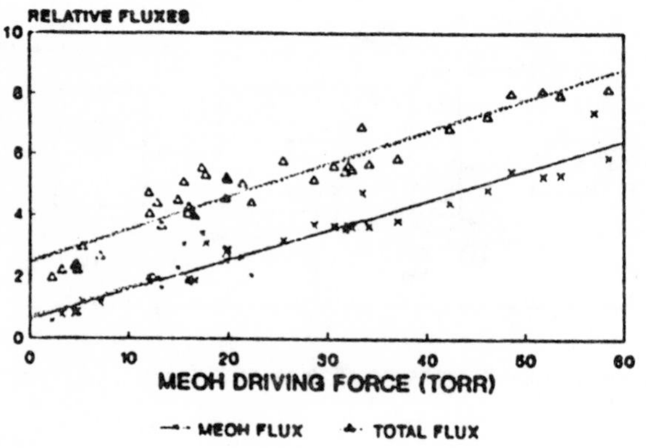

Figure 10. Demo test element fluxes.

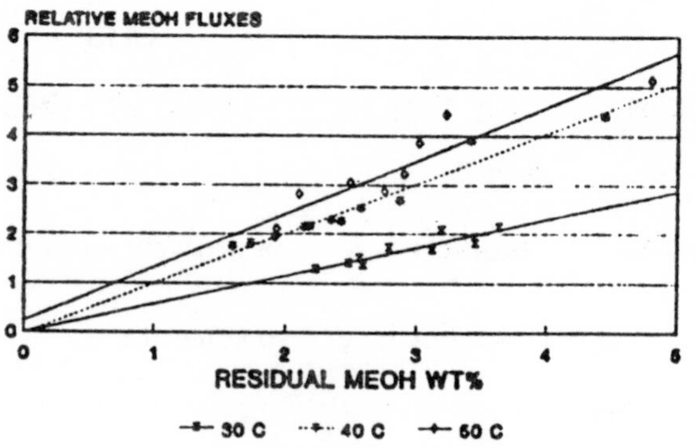

Figure 13. Effect of temperature demonstration skid data.

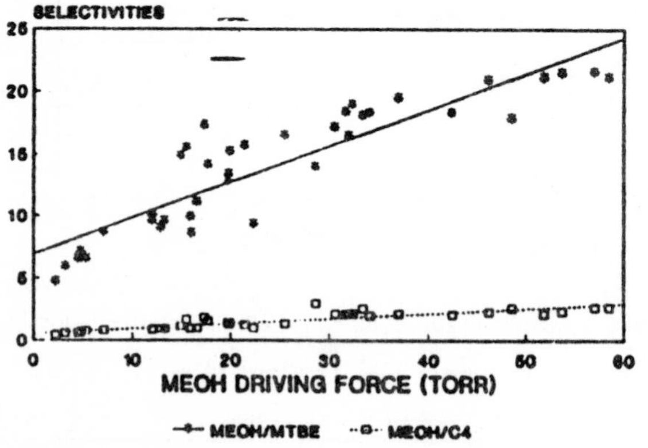

Figure 11. Demo test element selectivities.

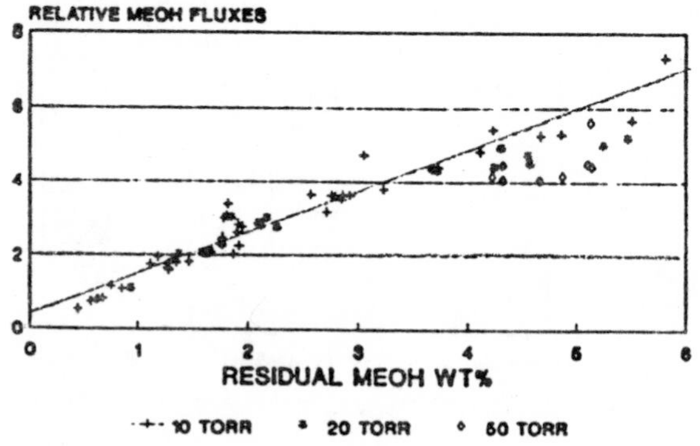

Figure 14. Effect of vacuum level demonstration skid data.

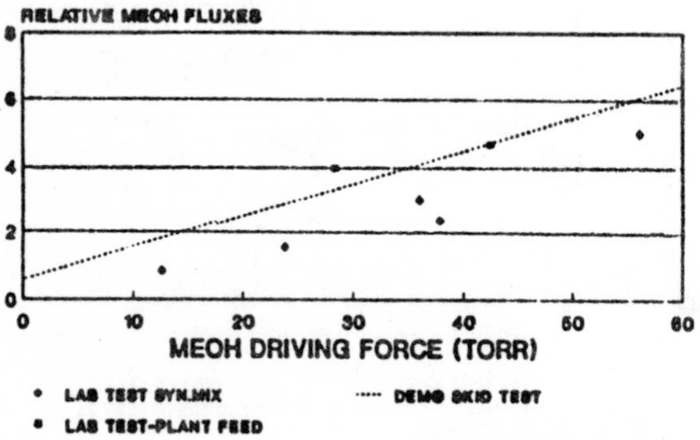

Figure 12. Demo skid element performance comparison with lab data.

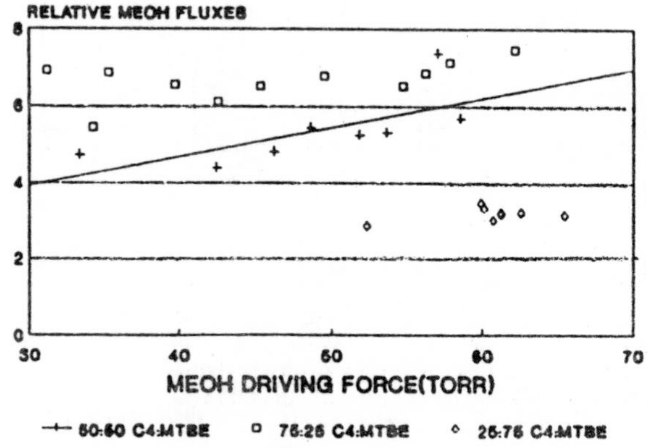

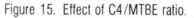

Figure 15. Effect of C4/MTBE ratio.

PERVAPORATION AS AN ALTERNATIVE PROCESS FOR THE SEPARATION OF METHANOL FROM C_4 HYDROCARBONS IN THE PRODUCTION OF MTBE AND TAME

Brian Anthony Farnand ■ Energy Mines and Resources Canada (CANMET), 555 Booth Street, Ottawa, Ontario, Canada
Soo Hong Noh ■ Zenon Environmental Inc., 845 Harrington Court, Burlington, Ontario, Canada

Pervaporation was investigated as an alternative process for the separation of methanol from a mixture of C_4 and C_5 hydrocarbons and methyl ethers in the production of MTBE (methyl tertiary-butyl ether) and TAME (tertiary-amyl methyl ether). A preliminary screening of pervaporation membranes was performed by using reactor effluent produced in a pilot scale etherification reactor. Nafion and a cellulose based membrane were chosen for further testing and showed similar performance.

INTRODUCTION

In response to the phase out of leaded components in gasoline, alternative sources of octane improvers are required. These include aromatics, methanol and oxygenates, as well as hydrocarbons produced from more typical refinery operations such as isomerization, alkylation, and polymer gasoline. In particular, the light hydrocarbon olefins in the C_4 streams have been useful, since they are produced within a refinery. Further, these olefins are produced to a greater extent in the present octane "squeeze" conditions as refiners operate their fluidized catalytic cracking units at higher severity.

The competition inside the refinery for these olefins is critical to the introduction of new processes for octane improvement. As an example, MTBE uses some of the olefins that would normally be used for alkylation. The competition for these olefins is further complicated by the sensitivity of alkylation catalysts (hydrofluoric acid and sulphuric acid) and by the presence of oxygenates such as alcohol and water. Only isobutylene reacts to make MTBE, while all the other olefins react to a significant extent in alkylation. Present refinery operation with higher severity to increase the octane content produces more reactive olefins than existing alkylation units can process efficiently. This increased supply of olefins influences both the economics and the technical feasibility of introducing new MTBE units into refineries that have existing alkylation units or other C_4 olefins.

The reaction to produce methyl ethers for gasoline octane improvement requires iso-olefins and methanol. Unreacted methanol and the ether product must be removed from the reactor effluent to avoid poisoning alkylation catalysts and to reduce the vapour pressure of gasoline, which is adversely affected by the thermodynamic non-ideal behaviour of methanol in hydrocarbons (Chase and Galvez, 1981; Unzelman, 1984). This nonideal behaviour results in the formation of a vapour-liquid azeotrope that prevents the use of distillation. A significant portion of the cost of ether manufacture is directly attributed to the removal of oxygenates from the reactor product, and the yields of ether are limited by the low methanol concentrations used for the ease of removal. These costs have inhibited the manufacture of MTBE in refineries and the usual source in the North American fuels refinery is from integrated petrochemical operations outside the refinery limits, where pure isobutylene streams can be dedicated to MTBE manufacture, and unreacted methanol in the reactor product can be recycled to extinction (Bitar et al., 1984). Thus a potential source of inexpensive isobutylene produced inside the refinery is not exploited for high octane blending components.

The purpose of this study is to investigate the removal of unreacted methanol from etherification reactor effluent for a refinery operation. Despite the previous work using reverse osmosis for this separation (Farnand and Sawatzky, 1986), the use of pervaporation has been chosen because of the high selectivity for the removal of methanol from hydrocarbons. Pervaporation does not introduce extraction solvents into the refinery and is not a cyclic operation such as adsorption. Operation at more advantageous concentrations for MTBE production could be possible if the separation of methanol could be performed efficiently (Smith and Huddleston, 1982). Further, the effective removal of methanol from the reactor effluent, followed by distillation to remove the ether product would permit the effective operation of both etherification and alkylation reactors on the same original feed of olefin rich hydrocarbons. Other methods that have been studied for this separation include water washing, glycol extraction, molecular sieve adsorption, and reverse osmosis.

EXPERIMENTAL

A schematic diagram pervaporation equipment used for this experiment is shown in Figure 1. The temperature of the experiment was controlled by immersing the pervaporation test cell in a controlled temperature water bath. Two cold traps were used for collecting the membrane permeate to permit steady state to be achieved. The membrane test cell used did not include a sweep gas for the low pressure side. The high pressure side of the apparatus was slightly pressurized to prevent losses by evaporation. Samples of etherification reactor effluent were supplied by PetroCanada Products, who also performed the chemical analyses.

RESULTS AND DISCUSSION

A survey of commercially available thin films was made in an effort to determine quickly which polymeric materials would give selective separation of the methanol or the hydrocarbons. This is similar to the approach used to select reverse osmosis membrane to perform the same separation. First, an attempt was made to measure the contact angle of methanol with the surface of the membrane to give an approximate determination if the membrane could be considered to be methanol attracting or rejecting, as reported in Table 1. These were measured by placing a drop of methanol on the surface of the membrane and measuring the contact angle with a microscope.

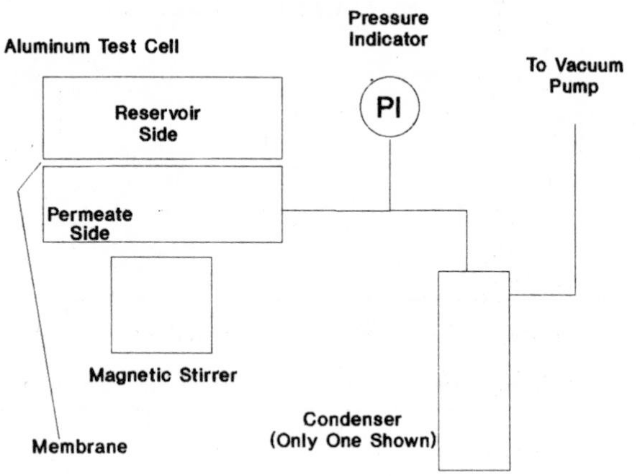

Figure 1 Pervaporation Test Cell Schematic

Table 1 Properties of Selected Membrane Materials

Membrane	Membrane Material	Thickness μm	Contact Angle
Kodacel TA	Cellulose Esters	76	62°
PVA	Polyvinyl Alcohol	76	-
MT	Cellulose Base	76	69°
Ultem	Polyetherimide	127	80°
Stabar K200	Polyether Ketone	25	86°
Stabar S100	Polyether Sulphone	25	85°
Nafion 117	Perfluorinated Copolymer	178	-
Nafion 417	Perfluorinated Copolymer	254	-
Lexan	Polycarbonate	51	85°
Saran F-120	Vinylidene chloride-acyrlonitrile	25	-
Kynar	Polyvinylidene fluoride	51	89°
UWB	Polyacrylic Acid-Nylon 6	-	-

The thickness of the membrane was measured and is also shown in Table 1. Those films which were not completely wetted by the methanol were eliminated from further study, since they probably would not selectively permeate methanol as well as the methanol wetted films.

Initial membrane screening experiments were performed with the methanol wetted membranes with a solution of 10% methanol in pentane. These results are reported in Table 2. There is no obvious relation observed between methanol contact angle and the pervaporation performance reported in this work. The two membranes chosen for further investigation are Nafion 117, a polymeric perfluorosulfonic acid typically used for electrodialysis, and MT, a cellulose based commercial thin film packaging barrier. These two membranes were then tested in pervaporation experiments to assess their performance as a function of methanol concentration in a methanol-pentane solution. The variation of permeation rate and methanol separation factor are shown in Figures 2 and 3. It is apparent that as methanol concentration decreases, the selectivity for methanol increases and the permeation rate decreases.

Table 2 Membrane Screening Experiment Results[a]

Membrane	Permeation Rate (mL m^{-2} h^{-1})	Permeate MeOH Conc Vol %	Separation Factor
Nafion 117	323.1	83	44
PVA	104.9	87	60
Saran F120	b	-	-
Stabar K200	c	-	-
Stabar S100	15.6	83	44
MT	260.5	89	73
Ultem	b	-	-
UWB	66.3	85	51

[a] Feed methanol concentration of 10% in pentane, room temperature, 40 mm of Hg.
[b] No visible permeate.
[c] Insufficient permeate for analysis.

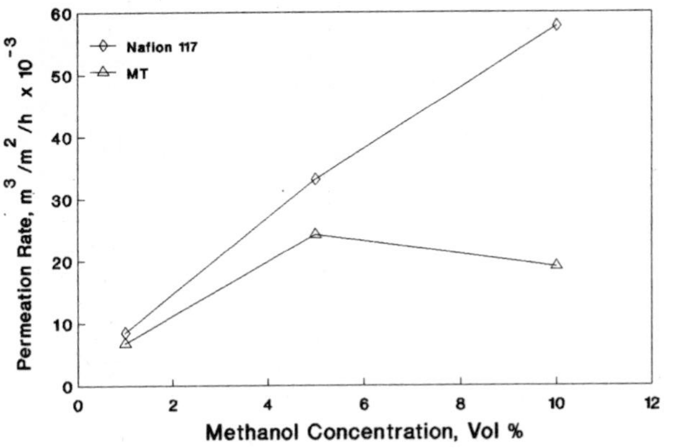

Figure 2 Permeation Rate With Concentration

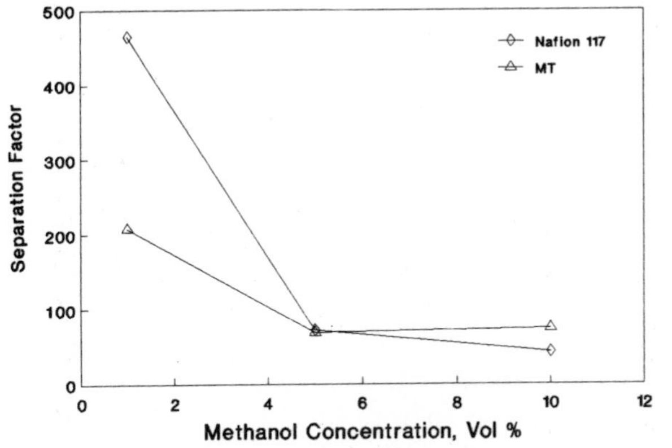

Figure 3 Separation Factor With Concentration

Experimental results with pilot plant generated etherification reactor effluent were also investigated. The composition of the two reactor effluents is shown in Table 3. The results of these experiments are shown in Tables 4 and 5. Poly(vinyl alcohol) membranes were included in these experiments for comparison with commercially popular membrane materials. It should be noted that the poly(vinyl alcohol) membranes used in this work were dense films and not the composite membranes used in commercial operations. The methanol concentration in the TAME experiments

Table 3 MTBE and TAME Reactor Effluent Concentration[a]

Component	MTBE	TAME
Methanol	3-5%	0.3-2%
MTBE	70-80%	0%
TAME	1-2%	10-20%
Isopentane	4-6%	30-40%

[a] Major components only.

Table 4 MTBE Reactor Effluent Pervaporation Experimental Results

Membrane	Permeation Rate mL m^{-2} h^{-1}	Permeate MeOH Conc Vol %	Separation Factor
Nafion 117	41	58.5%	35
Nafion 417	145	50.2%	25
Nafion 417	490	58.0%	35[c]
PVA	b	-	-
MT	89	55.6%	32
MT	102	71.4%	63[c]

[a] 40 mm Hg, Room Temperature unless noted, feed concentration as in Table 3.
[b] No permeate.
[c] 50°C.

Table 5 TAME Reactor Effluent Pervaporation Experimental Results[a]

Membrane	Permeation Rate mL m^{-2} h^{-1}	Permeate MeOH Conc Vol %	Separation Factor
Nafion 117	54	98.5%	3283
Nafion 117	314	89.7%	445[c]
PVA	b	-	-
MT	30	90.6%	3002[d]

[a] 40 mm Hg, Room Temperature unless noted, feed concentration as in Table 3.
[b] No permeate.
[c] 50°C.
[d] 45°C.

was lower than for the MTBE experiments since the processing strategy was to use a very low methanol content in the reactor. The olefins contained in the distillation fraction used for TAME production is usually not processed elsewhere in a refinery, so that there is no competition for the olefins. The TAME reactor effluent can be blended directly into the gasoline pool providing the vapour pressure increase caused by the addition of methanol is not large though the methanol must be matched with a cosolvent to prevent cold weather methanol phase separation. A successful methanol removal process applied to the TAME effluent stream would permit the use of higher concentrations of methanol in the reactor, with a subsequent greater octane increase. Further, the specifications for methanol in the final gasoline product without the use of a co-solvent could also be used to advantage. As was seen with the methanol-pentane solutions, the permeation rates for the TAME solutions with lower methanol content were generally lower and the separation factors were higher than for the MTBE case. It should be noted that the poly(vinyl alcohol) membrane used did not have an observable permeation, but this is probably caused by the thickness of the membrane and its resistance to permeation.

CONCLUSION

Polar membranes can be used in pervaporation processes to selectively remove methanol from both binary solutions in pentane as well as from MTBE and TAME reactor effluent. This may be attractive for the removal of methanol from reactor effluent as the processing of methyl ethers for use in gasoline inside fuels refineries becomes more financially attractive. Low permeation rates at dilute methanol concentration indicate that complete removal of methanol is difficult to obtain by pervaporation with the membranes studied in this work, and will require a polishing step to protect downstream catalysts. Further work will investigate the use of continuous pervaporation as opposed to the static cells used in this work.

REFERENCES

Bitar, L.S., Haxbun, E.A., Piel, W.J., Hydrocarbon Processing, p. 93, October 1984.

Chase, J.D., Galvez, B.B., Hydrocarbon Processing, p.89, March, 1981.

Farnand, B.A., Sawatzky, H., in Proceedings of the International Membrane Conference on the 25th Anniversary of Membrane Research in Canada, M. Malaiyandi, O. Kutowy, F. Talbot, editors, September 24-26, 1986, p. 229.

Smith, L.A., Huddleston, M.N., Hydrocarbon Processing, p. 121, March, 1982.

Unzelman, G.H., Oil and Gas Journal, p. 59, July 2, 1984.

OPPORTUNITIES FOR MEMBRANES IN THE PRODUCTION OF OCTANE ENHANCERS

V.M. Shah, C.R. Bartels, M. Pasternak,. and J. Reale ■ Texaco Inc., P.O. Box 509, Beacon, NY 12508

The phaseout of lead from gasoline based fuels has created an increased need for alternative additives. In this study, dimethyl carbonate (DMC) was considered as an alternative to methyl tertbutyl ether, which is the most widely used octane enhancer today. In the production of DMC, a DMC/methanol azeotrope is formed. Two processes, viz., (i) high pressure distillation and (ii) hybrid membrane process (pervaporation followed by distillation), were considered for this azeotropic separation. The hybrid membrane process is less capital intensive and has lower utility costs.

INTRODUCTION

The phaseout of lead from gasoline-based fuels has created an increased need for alternative additives. Besides boosting the octane number of the fuel, the additive should not substantially increase the Reid vapor pressure (RVP), should have a good water resistance (phase enhancer), and should be non-toxic. The total oxygen content of the resultant gasoline fuel is also regulated. Methyl-t-butyl ether (MTBE) is the most widely used octane enhancer. Dimethyl carbonate (DMC) or DMC blend with alcohols or with MTBE could be considered as alternate gasoline additives. Both MTBE and DMC can also serve as phase enhancers in alcohol containing gasolines (gasohols). However, isobutylene, which is a raw material for producing MTBE, is and has been a historically critical raw material. In addition, the price of MTBE is sensitive to the crude oil cost. On the other hand, DMC price sensitivity is related to methanol (MeOH) which could be coal based. Although, MTBE is less expensive to produce at current oil prices, this situation could reverse itself when the oil prices increase.

In a proprietary process for the production of DMC, a DMC/MeOH azeotrope is formed at atmospheric pressure. This azeotrope is separated to produce high purity (99 wt.%) DMC and the MeOH is recycled to the reactor. For optimum reaction efficiency, it is desirable to have no less than 95% MeOH in the recycle. It can be observed from the vapor-liquid equilibrium (VLE) curve (Figure 1) that at atmospheric pressure DMC and MeOH form an azeotrope at a concentration of 69% MeOH. At higher pressures this equilibrium shifts to higher MeOH concentrations. For example, at a pressure of 1172 kPa (170 psi) the azeotrope shifts to a concentration of 96% MeOH and at a pressure of 1724 kPa (250 psi) the DMC/MeOH system does not form an azeotrope. Two alternate processes, viz., (i) high pressure distillation and (ii) hybrid membrane (membrane assisted distillation) process, were considered for separating this azeotrope.

In the hybrid membrane process, membranes were used to just break the azeotrope. A reference to Figure 1 indicates that the K (=y/x) values are high at below-azeotropic MeOH concentrations. Hence, the remaining separation can be easily performed with a distillation column. The following section discusses the membrane performance for the desired separation. The subsequent section discusses the relative economics of the two separation processes.

MEMBRANES FOR DMC/MEOH SEPARATION

Pervaporation was used to separate the DMC/MeOH azeotrope. In the pervaporation process the liquid feed species dissolves into the membrane, diffuses through it, and then undergoes a phase change to a vapor at the back surface of the membrane. The permeate vapor is recovered by condensing at a low temperature. The permeate side of the membrane is kept under a vacuum, which is partially created by condensation of the permeate vapors. Since, the vapor is removed from the permeate side of the membrane, there is a concentration gradient across the membrane, which provides the driving force for permeation.

Various composite membranes were synthesized for this application. The porous support matrix of a Texaco composite membrane was made from poly (acrylonitrile) (PAN). The non-porous separating layer on top of the PAN was formed from poly(vinylalcohol) (PVA) crosslinked with a aliphatic dialdehyde. The cross-linking was carried out in the presence of an acid catalyst. The membranes were then cured at a high temperature. For comparison, commercial membranes also made of a PVA separating layer on a PAN support matrix were evaluated.

The Texaco and the commercial membranes were evaluated at various concentrations of the DMC/MeOH mixture. Representative results are presented in Table 1. It can be observed from this table that the Texaco membrane had higher flux <u>and</u> a lower permeate DMC concentration than the commercial membrane. Hence, the results of the Texaco membrane were used to compare the economics of a hybrid membrane process with high pressure distillation.

ECONOMIC EVALUATION

A DMC production rate of 907 metric tons (2 million lbs) per year was assumed. The two processes and their relative economics are discussed below:

High Pressure Distillation

The simulation of the high pressure distillation process was performed at 1724 kPa. 99% DMC was obtained from the bottom of the column and 95% MeOH from the top. The separation required thirty theoretical trays and the reboiler duty was 0.82 MW (2.8 MM Btu/hr). High pressure steam was used as the heating medium in the reboiler. The capital and utility costs are listed in Table 2.

Hybrid Membrane Process

The hybrid membrane process in this case denotes a combination of a membrane process and distillation for the desired separation. As shown in Figure 2, the azeotrope is first fed to a membrane unit. The permeate, containing 95% MeOH, is recycled to the reactor. The retentate, which contains 55% MeOH, is distilled at a pressure of 310 kPa (45 psi) to obtain 99% DMC as bottoms product. The overhead product, containing azeotropic composition of DMC and MeOH, is recycled to the membrane unit.

It can be observed form Table 1 that both, the flux through the membrane and the permeate concentration of DMC increased with an increase in feed temperature. Assuming that the membrane unit was operated at about 75 to 80°C, the permeate flux was calculated to be 0.6 kg/(m^2 hr) (kmh). Hence, a pervaporation unit with a membrane area of 540 m^2 is required. The permeate vapors were recovered by condensing and cooling them to a temperature of 0°C. The separation of the retentate required a distillation column with nine theoretical trays and a reboiler duty of 0.20 MW (0.68 MM Btu/hr). Medium pressure steam was used as a heating medium in the reboiler. The capital and utility costs of this membrane assisted distillation process are listed in Table 2.

CONCLUSIONS

On the basis of preliminary economic calculations, the capital cost for high pressure distillation is 50% higher than that for a hybrid membrane process. Since, a membrane process is modular in nature this difference would be reduced for larger size plants. The membrane process, however, has a distinct advantage in terms of utility costs. This is because high pressure distillation utilizes high pressure steam which is expensive. On the other hand, the membrane process requires low pressure (or waste) and medium pressure steam. Secondly, the total amount of steam required in the hybrid membrane process is lower than that required in high pressure distillation. For both processes, the electricity and cooling water costs are small compared to the steam costs.

Table 1

Comparison of Texaco and Commercial Membranes for DMC/MeOH Separation

Membrane	Temp. (°C)	30% DMC in Feed: Flux (kmh)*	30% DMC in Feed: Perm. Conc. (Wt.% DMC)	45% DMC in Feed: Flux (kmh)*	45% DMC in Feed: Perm. Conc. (Wt.% DMC)
Commercial	60	0.02	4.5	0.01	14.2
Texaco	60	0.12	1.7	-	-
	70	0.34	2.9	0.27	3.8

*kmh=kg/(m^2 hr)

Table 2

Cost Comparison Between High Pressure Distillation and Hybrid Membrane Process for DMC/MeOH Azeotrope Separation

DMC Production Rate: 907 metric tons/year

	High Pressure Distillation	Membrane Assisted Distillation
CAPITAL COST ($,x10^6)	1.5	1.0
Annual Utility Requirements		
1. Steam (x10^6 kgs)		
a. Memb. Unit (138 kPa)	-	1.1
b. Distillation Column		
i) 1.38 MPa	-	2.9
ii) 4.14 MPa	14	-
2. Electricity(x10^9 kJ) for refrigeration	-	1.1
3. Cooling Water(x10^6 m^3)	0.3	0.1
	-----	-----
UTILITY COST ($, 000)	171	45
	-----	-----

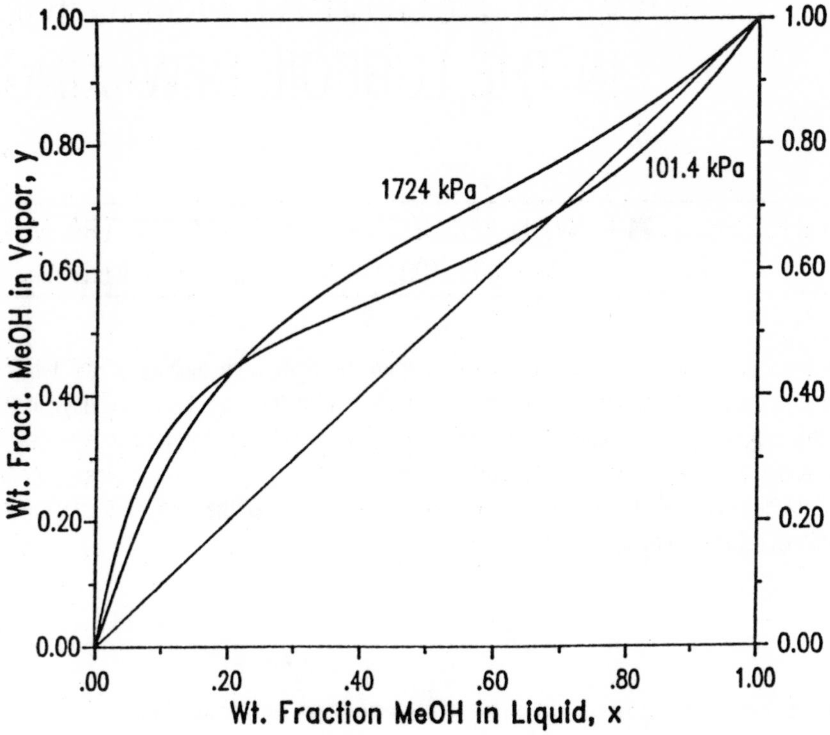

Figure 1. Vapor liquid equilibrium diagram for DMC/MeOH.

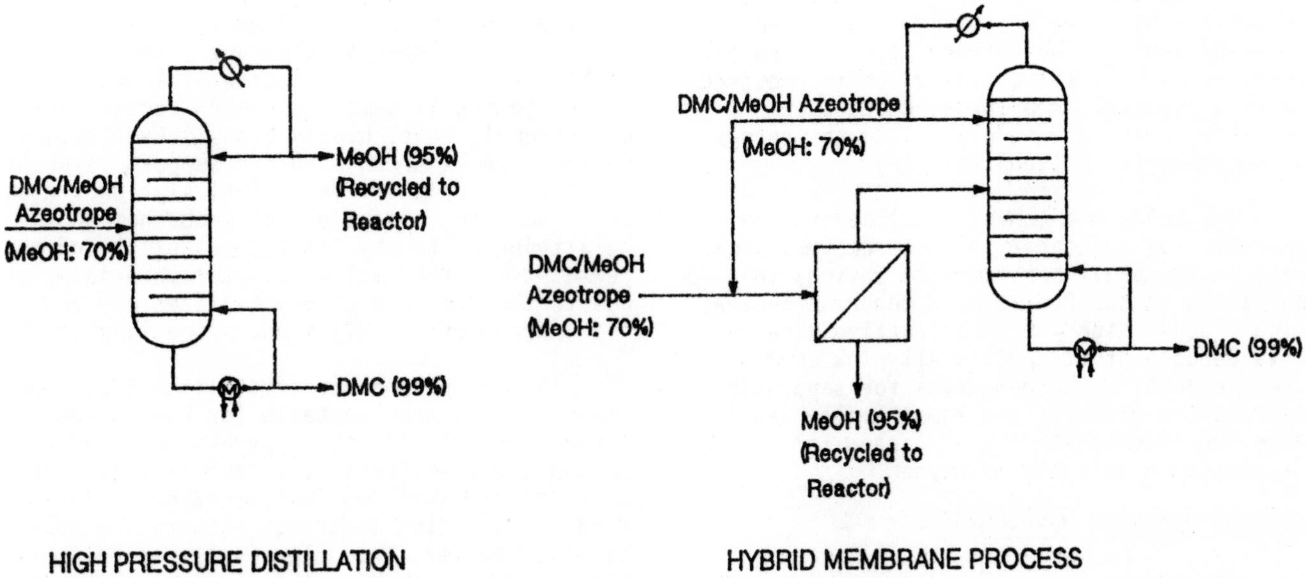

Figure 2. Process configurations for DMC/MeOH azeotrope separation via (i) high pressure distillation and (ii) hybrid membrane process.

SOLVENT RECOVERY USING MEMBRANES IN THE LUBEOIL DEWAXING PROCESS

J.G.A. Bitter, J.P. Haan and H.C. Rijkens ■ KONINKLIJKE/SHELL-LABORATORIUM, AMSTERDAM, (Shell Research B.V.), P.O. Box 3003, 1003 AA Amsterdam, The Netherlands

A novel membrane has been developed for separating solvents from the oil/solvent mixture of the luboil dewaxing process. The membrane, which consists of a modified silicone rubber, shows a high flux (more than 1 $m^3/(m^2.day)$ at 40 bar) and a selectivity varying between 50 and 100, depending on solvent dilution and grade of luboil. The membrane has now been in continuous operation on a laboratory scale for about four years, without showing a significant change of its properties. Economic evaluations indicate that the operation costs (including capital charges) of a membrane unit based on the use of this membrane are lower than the energy costs of the conventional separation by flashing.

Lubrication base oils contain waxes, which solidify upon cooling down. As these waxes would cause a bad cold start of motorcars in wintertime, they are separated in a so-called solvent dewaxing plant. In this process the oil feed is diluted with an organic solvent mixture, after which the wax is caused to crystallize by cooling the mixture down to -20 ^{o}C and is then filtered off. Next the solvent mixture is separated from the dewaxed oil by flashing and stripping in a solvent recovery section (see Figure 1). For a smooth operation a high solvent dilution is required in this process. As a consequence, the operating costs are mainly due to the energy consumption for solvent recovery.

Replacing the conventional separation procedure by a cheaper solvent recovery process would entail considerable savings in operating costs. Possibly, a membrane separation process might be an attractive alternative because of its potentially low energy requirements. As no membranes for separating hydrocarbon mixtures are commercially available, an investigation was started for developing a suitable membrane.

SOLVENT RECOVERY MEMBRANE

For separations on a molecular scale so-called tight (diffusion-type) membranes must be used. In such membranes the permeants dissolve in the membrane material and pass by diffusion in response to their concentration gradient inside the membrane, caused by a driving force exerted externally.

For the separation of aqueous mixtures (solutions) membranes are commercially available. Most of these membranes are semi-crystalline polymers. Because transport takes place only in the amorphous polymer phase and not in the crystallites, the permeation rate is very low in such materials. In order to achieve reasonable fluxes asymmetric membranes are made, which consist of an extremely thin and tight top layer on a porous support layer of the same polymer. Another well-known type is the composite membrane, which consists of an extremely thin, highly crosslinked (resin) polymer top layer on a porous support layer of a different polymer. The relatively low fluxes observed with these types of membranes should be attributed to the limited solubility of the permeants in the applied polymer materials, due to the presence of crystallites and the high degree of crosslinking, respectively.

In our opinion non-crystalline elastomers (rubbers) are more suitable for use as membrane materials for the separation of hydrocarbon mixtures. Generally such materials show considerable swelling in hydrocarbons, resulting in high flux membranes of very low selectivity, however. The selectivity can be improved either by altering the chemical structure of the polymer or by increasing its degree of crosslinking. For instance, a polydimethylsiloxane (pdms) membrane showed a swelling (solvent absorption) of more than 300 %v in a mixture of methyl ethyl ketone, toluene and a residual luboil fraction (volume ratio 1:1:1). The flux through this membrane amounted to more than 4 $m^3/(m^2.day)$ at 40 bar permeation pressure; the selectivity of the solvent mixture with respect to the luboil was less than 7. For a fluorinated siloxane polymer the swelling in the same hydrocarbon mixture reduced to 100 %v. The lower swelling of this fluorosilicone rubber is caused by a much lower solubility of toluene (less than 30 %v versus more than 800 %v in pdms), which now acts as a deswelling agent for the ketone. The fluorosilicone membrane showed at corresponding conditions a flux of about 1.5 $m^3/(m^2.day)$ at 40 bar and a selectivity of more than 50 with the luboil/solvent mixture previously mentioned. Other experimental data are collected in Figures 2 and 3, which show the large effect of solvent dilution of the feed mixture on permeate flux and selectivity, respectively.

In a laboratory experiment a fluorosilicone membrane has now been in continuous operation with an oil/solvent mixture for more than four years without showing a significant alteration in properties.

MEMBRANE SEPARATION PROCESS

In principle two types of membrane separation processes may be considered for solvent recovery, viz. dialysis and reverse osmosis. In the former process a diluent is applied on the permeate side of the membrane for reducing the permeant concentration, thus creating a concentration gradient across the membrane and causing transport of the permeant. Fresh (untreated) waxy feed can be used as a "diluent" for the solvent from the oil/solvent mixture. In this way solvent transfer from the treated to the untreated feedstock takes place via the membrane, thus replacing a major part of solvent flashing. This process is probably the most economical one from a viewpoint of energy saving because no energy is required for the solvent transfer. However, for implementation in an existing solvent dewaxing plant major alterations are required. Furthermore, no technically viable dialysis module is currently commercially available.

Alternatively reverse osmosis may be applied. In this process solvent transport takes place under the influence of a pressure drop maintained across the membrane. This process consumes energy, although much less than the conventional solvent recovery processes. Because the reverse osmosis unit can be installed parallel to the solvent recovery unit it is also suitable for existing plants. In reverse osmosis solvent separation is limited by the occurring osmotic pressure. As shown in Figure 2 the optimum concentration achievable with residual luboil at 40 bar is about 30 %w. Therefore, starting from a 15 %w dewaxed oil mixture separation of at least 70 % of the solvent by reverse osmosis is technically feasible.

ECONOMIC EVALUATION

Based on the process schemes of Figures 1 and 4 an economical evaluation was made, using the experimental data shown in Figures 2 and 3. For the capital and membrane-replacement costs the data of existing membrane units for water purification were used. From the results in Table 1 it follows that even at the current low energy prices, the costs of installation and operation of a reverse osmosis plant for solvent recovery in existing solvent dewaxing units are already met by the savings in energy costs. The economic incentive increases considerably in the case of debottlenecking existing solvent dewaxing units or building new ones.

Table 1. Estimated costs (US$) per tonne of solvent

Flashing:	
Energy (US$ 85 to 200/t fuel)	0.58-1.46
Membrane separation:	
Energy	0.12-0.29
Capital	0.19
Membrane replacement	0.10
Labour	0.11
Materials	0.05
Total	0.57-0.74

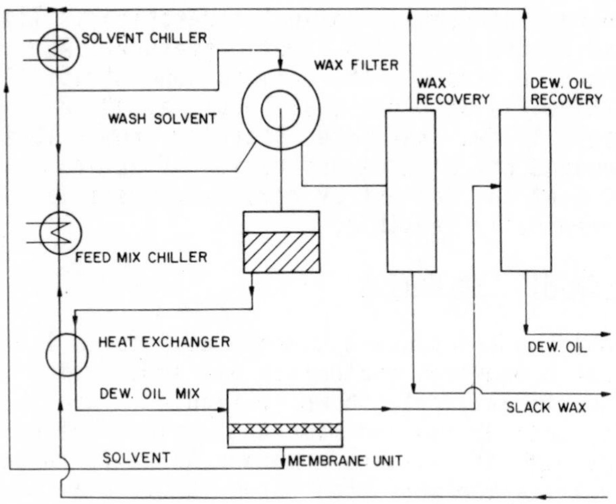

Figure 1. Solvent recovery by membrane separation

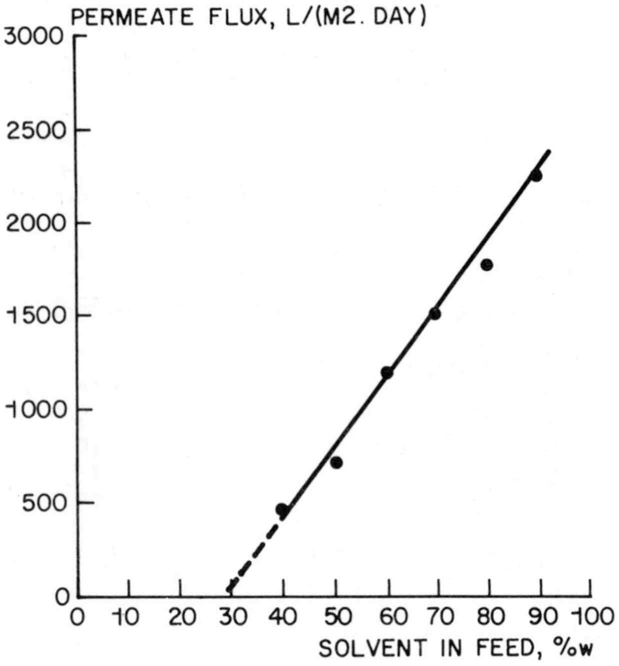

Figure 2. Fluorosilicone membrane, residual luboil fraction

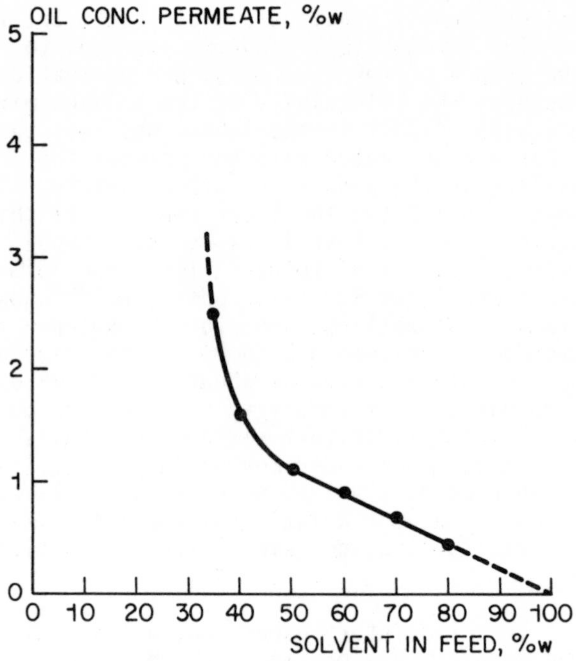

Figure 3. Fluorosilicone membrane, residual luboil fraction

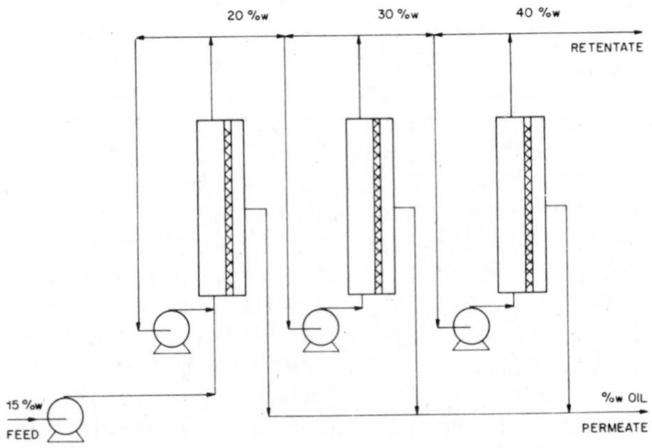

Figure 4. Membrane separation scheme for solvent recovery

PROCESSING OF CRUDE OILS WITH POLYMERIC ULTRAFILTRATION MEMBRANES

J.D. Hazlett, O. Kutowy, and T.A. Tweddle ■ National Research Council of Canada, Ottawa, Ontario, Canada
B.A. Fernand ■ Energy, Mines, and Resources Canada, (CANMET). Ottawa, Ontario, Canada

A variety of commercial and NRC developed polysulfone membranes were evaluated for the processing of crude oil feedstocks. The products of the process are a permeate which is the equivalent of a gas oil and a concentrate enriched in asphaltenes. A systematic study of the effects of temperature, pressure and circulation flow has been completed for Mydale crude using a plate and frame membrane module. The establishment of a gel layer on the membrane surface has been demonstrated. The influence of this dynamic gel-polysulfone composite membrane on the flux and separation characteristics is discussed.

INTRODUCTION

Membrane separations using synthetic polymeric membranes are well established for water purification and aqueous applications in the areas of food processing and biotechnology. Non-aqueous applications of membrane technology are a more recent development. In earlier papers (1,2), the application of membrane processing to the fractionation of conventional crude oils, heavy oils and solvent diluted bitumens has been discussed. Pure solvent experiments (3) suggested that the lower molecular cut-offs experienced in hydrocarbon service resulted from a tightening of the membrane pore structure when the polymeric membrane swelled. The present work describes a systematic investigation of the effect of operating parameters on process performance for one crude oil feed.

LITERATURE REVIEW

The focus of the majority of the published research has been either the separation of molecules of very different molecular size or the removal of particulates from waste solvents and oils. For instance, one of the earliest non-aqueous applications to be investigated was the regeneration of used automotive oils (4). A Japanese Company (5) developed tubular polyimide membranes for the recovery of solvents from waste paints and polymeric resins, and the removal of solids from waste oils.

Several patents have proposed membrane processes as alternatives in the production of lube oils. A two step approach involving solvent extraction to remove aromatics followed by dewaxing via a crystallization/filtration process is commonly employed to produce lube oil base stocks. Different solvents must be used in each of these steps to obtain the desired separations. An alternative to solvent extraction for the removal of aromatic compounds, involves the addition of traditional dewaxing solvents to the feed oil and recovery of an aromatics rich permeate using regenerated cellulose reverse osmosis membranes (6). The paraffinic retentate is processed using conventional dewaxing technology. The use of a second solvent has been eliminated by the introduction of a membrane process.

As an alternative to distillation, the recovery of dewaxing solvents by a membrane process has been proposed (7). The preferred membrane is a 1 to 10 μm layer of halogen substituted silicone compounds on a 15 to 100 μm polypropylene support. It was discovered that the combination of an aromatic solvent with a polar aliphatic solvent (e.g. toluene and methyl ethyl ketone) resulted in a relatively pure solvent permeate. For example, solvent recovered from dewaxed bright stock oil using three membrane separation stages, contained less than 1% of the hydrocarbon oil, making it suitable for recycle to the dewaxing unit without further processing.

Another possible non-aqueous membrane process is similar to dialysis (8). Improved efficiency in solvent extraction operations can be achieved by the use of membranes which allow the permeation of solvent from the separated extract to the feed. For example, in the case of heavy oil deasphalting, deasphalted oil extract passing on one side of a membrane will transfer solvent to pre-diluted feed passing by the other surface of the membrane. The solvent content of the extract is reduced resulting in lower solvent stripping costs. With the choice of certain membranes, a portion of the oil will pass directly from the feed stream to the extract, further improving the process efficiency. Many other extraction processes could benefit from this type of flowsheet modification.

The recovery of pentane deasphalting solvent from deasphalted oils using spiral wound membrane modules has been developed by an American firm (9,10,11). The formation of a gel layer, "a relatively thick layer on the surface of the membrane" was identified. This gel layer had a beneficial effect on selectivity, but it decreased permeation rates. Increasing hydrophillicity, for instance by sulfonation of the surface of a polysulfone membrane, was found to limit the flux decline resulting from the formation of this gel layer. A four step solvent pretreatment of the polysulfone membranes when moving from the aqueous membrane casting environment to the non-aqueous process conditions was important in achieving higher permeation rates.

EXPERIMENTAL

Membrane Formation

Laboratory cast polysulfone membranes were made from Union Carbide Udel P1700 (U), Radel R5000 (R) and Imperial Chemical Industries Victrex 5200P (V). The polymers were dissolved in commercial grade N-methyl pyrrolidinone (NMP) with polyvinyl pyrrolidone (PVP) as a pore regulating additive. Casting dopes from these constituents were hand cast under controlled conditions onto spun bonded polyester backing. The nascent membranes were then gelled with ice water. NMP AND PVP were extensively leached from the gelled membranes by further water washing at room temperature. The membranes were solvent exchanged to prepare them for non-aqueous service. The solvent exchange procedure consisted of successive washings with commercial 95% ethanol, absolute ethanol and hexanes. The membranes were loaded into the membrane module once the solvent exchange was completed.

Commercial flat sheet membranes from De Danske Sukkerfabrikker (DDS) were also used. These included polyvinylidene fluoride (FS), Udel polysulfone (GR) and sulfonated polysulfone (GS) membranes with molecular weight cut-offs of 6,000, 20,000 and 50,000 as determined in aqueous service.

Crude Oil Feed

A sample of Mydale crude oil was obtained from the Alberta Research Council's Oil Sands Sample Bank. This Saskatchewan crude oil was produced via secondary, water flood recovery methods. This oil was chosen so that results of the current study could be compared with earlier work carried out in this laboratory. Table 1 presents the pertinent characteristics of this oil.

Table 1

Mydale Crude Oil

°API	28.8
N (wt %)	0.19
S (wt %)	2.07
Ni (ppm)	17
V (ppm)	27

Testing Procedures

Membranes were tested in a batch operation using a plate and frame membrane module, DDS's Lab Module 20. A support plate separates each pair of membranes, with flow passing across the membrane surfaces and permeate collecting in the internal channels of the plate. The permeate from each membrane pair can be sampled individually allowing the evaluation of several different membranes simultaneously. This unit has the capability of testing up to twenty pairs of membranes in series when used with aqueous systems. When treating oils, the number of pairs to be tested was limited to a maximum of ten pairs. This was due to the pressure drop that developed when viscous oil flowed through the flow passages in the spacer plates which separate the membrane/support plate sandwiches. When operating at temperatures ranging from 50 to 80°C, thermal expansion of the stainless steel rod that passes through the center of the membrane stack, resulted in flexure of the support plates and membrane failure due to small rips that formed in the membranes. This problem was overcome by reducing the number of membrane pairs to be tested to approximately five. For these reasons, the system was operated with two Lab Modules in parallel as shown in Figure 1. Each membrane pair provides 0.03 m^2 of useable area.

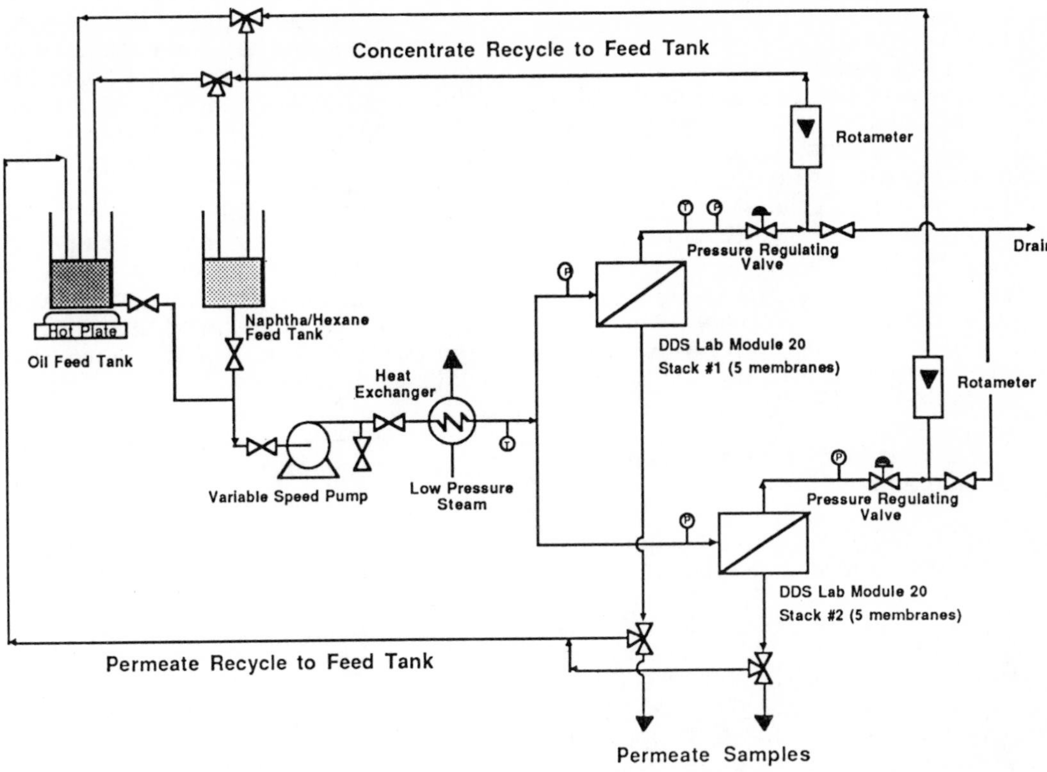

Figure 1: Schematic Diagram of Membrane Test System

Prior to introducing crude oil to the membrane system, the system was stored with hexanes filling the lines and covering the membranes. A known quantity of feed oil was placed in the feed tank at the start of the run. This feed material was heated using the hot plate while hexanes were circulated through the system. After approximately one hour, the pump was stopped and the hexanes drained from the system. The pump was restarted and crude oil feed was delivered under pressure to the membrane device where it separated into permeate and concentrate. During a period of approximately one hour, both permeate and concentrate were recycled to the feed tank. Final adjustment of the oil temperature was rapidly achieved with a low pressure steam heat exchanger. Once the desired operating temperature was reached, only the hot plate was required to maintain the system at a constant temperature. Three different types of experiments were performed:

1) Permeate and concentrate were continually returned to the feed tank, while the effect of operating pressure and recirculating flow rate on permeation flux and quality were determined.

2) Permeate and concentrate were continually returned to the feed tank while the effect of time on stream on permeation flux and quality were determined at fixed operating conditions.

3) Permeate was collected separately while the concentrate continued to recycle to the feed tank. Recovery as used in this paper is the percentage of oil originally charged to the feed tank collected as permeate. When a predetermined recovery was reached, the permeate was once again recycled to the feed tank while permeation flux and quality were measured under these new conditions.

The experiments described in this paper were carried out over a period of several weeks. Crude oil was never left in contact with the membranes unless there was circulation across the membrane surface to avoid irreversible fouling. The life of the membranes was significantly extended by draining the crude oil from the system at the end of each day, flushing the system with hexanes, and leaving the lines filled and membranes covered with clean hexanes.

Analysis of Permeate and Concentrate

Molecular weight averages and distributions were obtained using a Perkin Elmer Series 3B liquid chromatograph equipped with a differential refractometer detector. Three Waters Ultrastyragel GPC columns (1,000 Å, 500Å and 100Å) mounted in series were operated at 40°C using 1 ml/min of $CHCl_3$ as the eluting solvent. A sample injection volume of 100 µl was used. Oil samples were diluted prior to injection using $CHCl_3$ to provide loadings of ~7 mg/100 µl. Data was collected and processed using a Nelson Analytical 2600 Chromatography system. The HPLC system was calibrated using TSK polystyrene standards in $CHCl_3$.

Nickel and vanadium were analyzed using atomic absorption analysis.

Carbon-13 NMR measurements were performed on a Bruker MSL 300 spectrometer. A paramagnetic relaxation agent, chromium (III) acetylacetonate was added to the solutions of oil in deuterochloroform to allow for a short delay time between pulses in the Carbon-13 channel. Data was collected at a sweepwidth setting of 50 kHz with inverse gated proton decoupling to suppress nuclear Overhauser effect (nOe). A pulse flip angle of 45° was used with a pulse repetition time of 2 seconds. Between 10,000 and 20,000 transients were collected in order to obtain an adequate signal to noise ratio. Analysis of the resulting spectra established the percentage of aromatic carbon atoms and average chain length for the oil permeates.

DISCUSSION OF RESULTS

Effect of Temperature, Pressure and Recirculating Flow

Flux was measured at various pressures and recirculating flows with both permeate and concentrate recycling to the feed tank. A variety of flux-pressure relationships were observed between two extreme cases. In the case of the tight GR81 membrane (Figure 2), a linear relationship between flux and pressure was observed which was unaffected by increasing the recirculation flow. The more open V15-0 membrane (Figure 3), shows clear evidence of concentration polarization effects with flux approaching a limiting value as pressure is increased. The plateau in the flux-pressure relationship for this membrane occurs for pressures greater than 1.0 MPa. As the recirculation flow is increased from 5.7 to 9.5 l/min, the permeate flux increases. Further increase in this flow appears to have little effect on the measured flux. These observations suggest that the separation is the result of a dynamic gel-polysulfone composite. For tight membranes and for open membranes at lower pressures (<1.0 MPa), the limiting resistance is the inherent permeability of the polysulfone membrane. For open membranes at higher pressures, the polarized layer which is the result of the concentration profile near the membrane surface, becomes the limiting factor. An increase in recirculation flow, increases the turbulence in the vicinity of the membrane surface, altering the structure of this gel layer. Once a certain degree of turbulence is achieved, the effect of a further increase in flow is not clear from the data obtained to date.

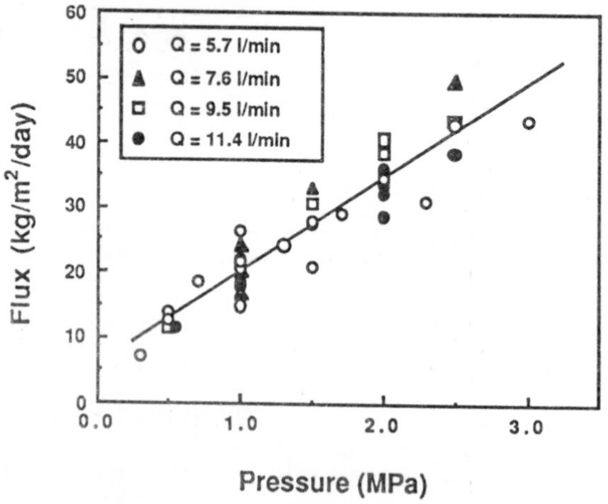

Figure 2: Variation of Flux with Pressure and Recirculation Flow for GR81 Membrane at 65°C

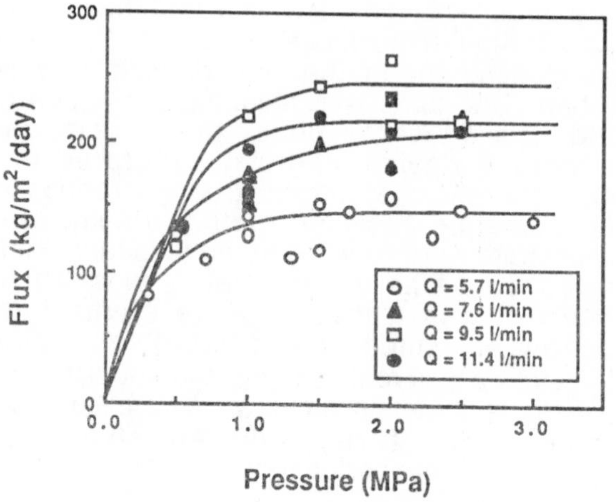

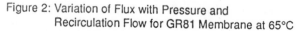

Figure 3: Variation of Flux with Pressure and Recirculation Flow for V15-0 Membrane at 65°C

It is of interest, that the quality (i.e. molecular weight distribution and demetallization) of permeate produced by membranes with molecular cut-offs ranging from

6,000 to 50,000 is strikingly similar. Figure 4 illustrates the molecular weight distributions for Mydale crude and gas oil permeate from a V15-8 membrane at 0% recovery. The higher molecular weight species that are present in the feed oil are clearly absent from the permeate. Generally speaking, the weight average molecular weight is the same for the permeate of all membranes tested. As can be seen from Table 2, the removal of contaminants such as Ni and V is similar for all membranes tested. The major difference between the various membranes is the measured permeation rate. It would appear that the membrane functions primarily as a surface on which the gel layer forms. The character of the separation appears to be controlled by this gel. As discussed previously, the flux can be limited by either the permeability of the membrane or transport through the gel layer to the membrane surface.

Table 2

Nickel and Vanadium Removal for Mydale Crude Permeates at 55°C, 1.5 MPa, Q = 5.7 l/min, 40% Recovery

Sample	Ni(ppm)	% Removal	V(ppm)	% Removal
Feed	29		53	
V15-8	3	90%	3	94%
V16-16	2	93%	3	94%
U16-16	2	93%	3	94%
V16-16	2	93%	3	94%
R16-16	2	93%	3	94%
FS60	3	90%	4	92%
GR51	2	93%	3	94%
GR61	2	93%	3	94%
GR81	2	93%	3	94%
GS61	2	93%	4	92%

Effect of Time on Stream

It was noted that time on stream affected the measured value of flux. Figure 5 presents the variation of permeate flux with time under a fixed set of operating conditions. As the flux declined, the character of the permeate was observed to change from a fluid liquid to a semi-solid. This decline is the result of two effects. Initially, the membranes are in mixed hexanes. When exposed to the more aromatic crude oil, the polymeric membranes swell, reducing the size of pores in the membrane and as a result, reducing the permeate flux. At the same time, the gel layer is beginning to form on the membrane surface. After 8 to 10 hours, the membrane flux

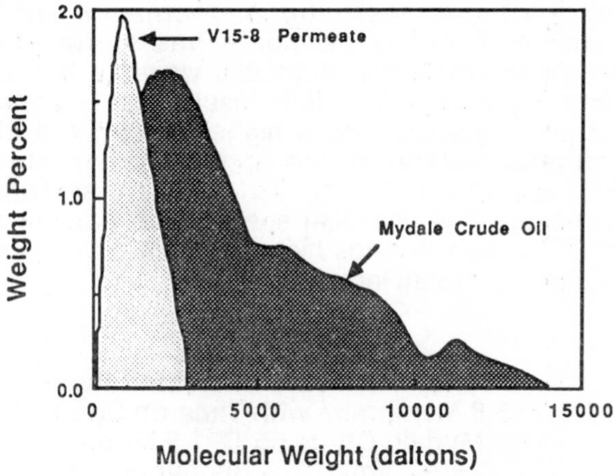

Figure 4: MW Curves for Crude Oil and V15-8 Permeate 55°C, 1.5 MPa, Q=5.7 l/min, 0% Recovery

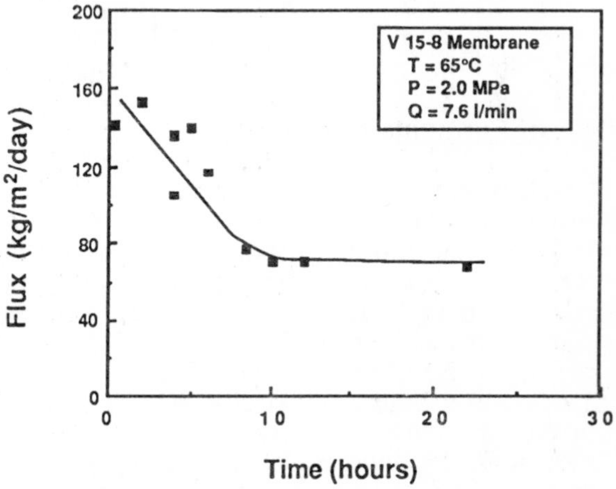

Figure 5: Variation of Flux with Time on Stream for V15-8 Membrane at 0% Recovery

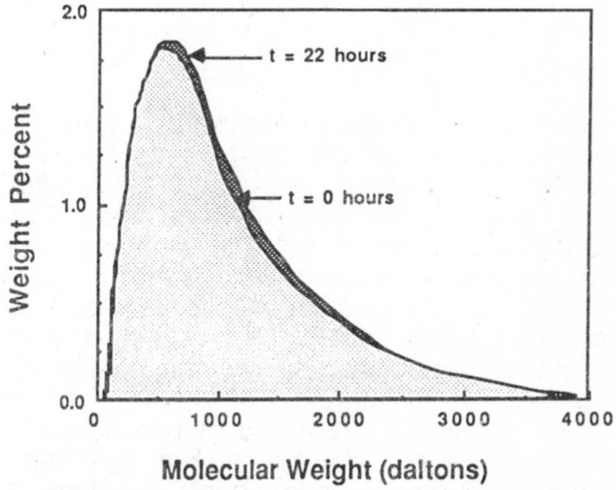

Figure 6: Variation in MW with Time for V15-8 Membrane 55°C, 1.5 MPa, Q=5.7 l/min, 0% Recovery

has stabilized. Figure 6 shows that during the 22 hours of this test, the molecular weight distribution of the permeate has shifted to include slightly higher molecular weight species. Table 3 presents C^{13} NMR results that indicate the permeate contains a higher percentage of aromatic carbon atoms and has a greater average chain length as time on stream increases. The chemical species that comprise this layer shift towards higher molecular weights as time on stream increases.

Table 3

Variation of Permeate Characteristics for V15-8 Membrane with Time on Stream for Mydale Crude 55°C, 1.5 MPa, Q = 5.7 l/min, 0 % Recovery

Time (hours)	C_{ar} (%)	L_c (carbon atoms)	$\overline{M}_w$ (daltons)
0	14.2	8.6	335
12	21.3	10.6	342
22	23.0	11.6	351

Effect of Permeate Recovery

When the system is operated without permeate recycle, the flux declines as recovery increases. This is to be expected as the viscosity of the feed material increases with increased recovery and the fraction of permeable species remaining in the feed is decreasing. The series of results at 55°C presented in Figure 7, show that operating below the region where the gel layer becomes the limiting resistance to flux (i.e. 0.5 MPa as opposed to 1.5 or 2.5 MPa in the present case), can extend the range of achievable recovery for a given temperature. It is also apparent that increased operating temperatures are effective in increasing the permeation rate. This is believed to be primarily due to the corresponding reduction in feed viscosity.

Figure 8 illustrates the shift in permeate molecular weight distribution as recovery increases. As lower molecular weight material is removed in the permeate, the feed material and the gel layer on the membrane surface become progressively heavier. This results in a shift towards higher molecular weight material in the permeate. This shift is also reflected in the weight average molecular weight as shown in Table 4. It is of interest that values of $\overline{M}_w$ for V15–8 permeate produced at two operating pressures are considerably different (see Tables 3 and 4). Operating under conditions where the gel layer is not limiting permeate flux results in a higher molecular weight permeate.

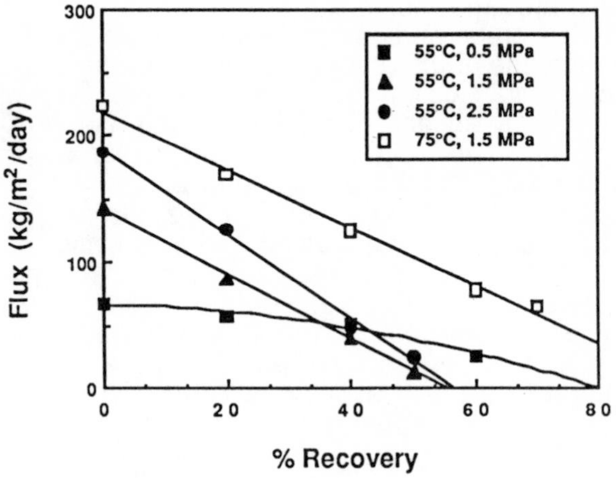

Figure 7: Variation of Flux with % Recovery for V15-8 Membrane, Q=5.7 l/min

Table 4

Variation of Permeate Molecular Weight for V15-8 Membrane with % Recovery for Mydale Crude at 55°C, 0.5 MPa, Q = 5.7 l/min

% Recovery	$\overline{M}_w$ (daltons)
0	558
20	573
40	600

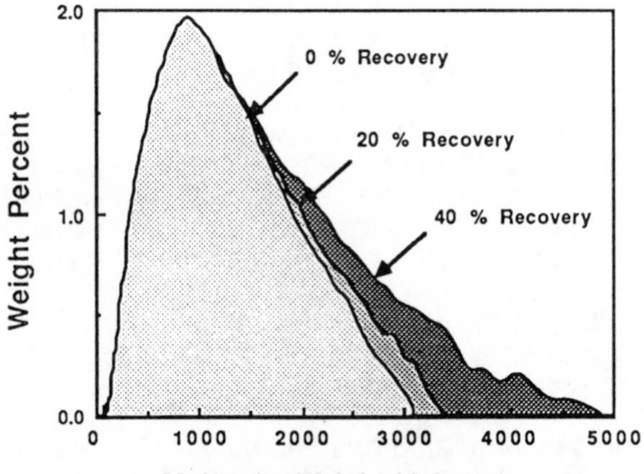

Figure 8: Variation of MW with % Recovery for V15-8 Permeate, 55°C, 0.5 MPa, Q=5.7 l/min

Operational Limitations

As mentioned earlier, limitations in the currently available membranes and membrane modules make operation at elevated temperatures difficult. For instance, polyethylene and polypropylene backings which are commonly used for commercial ultrafiltration membranes, are not stable in hydrocarbon service at temperatures approaching 80°C. Spun-bonded polyester has been used as the backing for the laboratory cast membranes to overcome this limitation. The combined effects of elevated temperature and hydrocarbon exposure caused failure of the polysulfone membrane support and spacer plates that are used in the DDS Lab Module 20. A new plate and frame module is needed to avoid this problem and the large pressure drops that result from the small flow passages in the spacer plates.

CONCLUSIONS

Polysulfone ultrafiltration membranes have been successfully used to fractionate crude oil. The separation appears to result from the formation of a dynamic gel-polysulfone composite membrane. Open membranes with an aqueous molecular weight cut-off of approximately 50,000 yield the highest permeate fluxes. Optimal operating conditions were found to be the maximum safe operating temperature (presently limited to ~70°C) and pressures in the vicinity of 1 MPa. While the removal of contaminants such as Ni and V was the same for different membranes, evidence suggests that the permeate molecular weight is influenced by operating pressure. Future work will focus on the role of membrane hydrophillicty on performance and the design of a plate and frame module that can withstand higher operating temperatures.

ACKNOWLEDGEMENTS

The significant contributions of M. Dal-Cin, R. Taticek and G. Clarkin, who carried out the experiments described here, are gratefully acknowledged. Many thanks to J. Woods and L. Kotylar for their assistance with HPLC and C^{13} NMR analysis.

This project was supported financially by CANMET/ERL, Energy, Mines and Resources Canada.

LITERATURE CITED

1. Kutowy, O., P. Guerin, T.A. Tweddle, J. Woods, "Use of Membranes for Oil Upgrading", Proc. 35th Can. Chem. Eng. Conf., **1**, 26-30, CSChE, Ottawa (1985).

2. Sparks, B.D., J.D. Hazlett, O. Kutowy, T.A. Tweddle, "Upgrading of Solvent Extracted Athabasca Bitumen by Membrane Ultrafiltration", presented at AIChE Summer National Meeting, Minneapolis, Minnesota, August 16-19, 1987.

3. Hazlett, J., O. Kutowy, T.A. Tweddle, M.D. Guiver, T.W. McCracken, "Polysulfone Membranes in Non-Aqueous Applications", Proceedings of the International Membrane Conference, 241-257, NRC, Ottawa (1986).

4. Parc, G., M. Born, A. Rojey, US Patent 3, 919,075, November 11, 1975.

5. Iwama, A., Y. Kazuse, J. Memb. Sci., **11**, 297-309 (1982).

6. Gudelis, D.A., M.M. Hafez, H.W. Pauls, D.H. Shaw, European Patent Application, EP 0 160 142, November 6, 1985.

7. Bitter, J.G.A., J.P. Haan, H.C. Rijkens, European Patent Application, EP 0 220 753, May 6, 1987.

8. Bitter, J.G.A., UK Patent, GB 2,116,071B, January 30, 1985.

9. Kulkarni, S.S., E.W. Funk, N.N. Li, "Hydrocarbon Separations with Polymeric Membranes", AIChE Symp. Ser., No. 250, Vol. 82, 78-84 (1986).

10. Chang, Y.A., S.S. Kulkarni, E.W. Funk, US Patent 4, 595,507, June 17, 1986.

11. Funk, E.W., S.S. Kulkarni, and Y.A. Chang, US Patent 4,617,126, October 14, 1986.

INDEX

SYMPOSIUM SERIES

ADSORPTION

96 Developments in Physical Adsorption
117 Adsorption Technology
219 Recent Advances in Adsorption and Ion Exchange
230 Adsorption and Ion Exchange—'83
233 Adsorption and Ion Exchange—Progress and Future Prospects
242 Adsorption and Ion Exchange: Recent Developments
264 Adsorption and Ion Exchange: Fundaments and Applications

AEROSPACE

33 Rocket and Missile Technology
52 Chemical Engineering Techniques in Aerospace

BIOENGINEERING

69 Bioengineering and Food Processing
84 The Artificial Kidney
86 Bioengineering ... Food
93 Engineering of Unconventional Protein Production
99 Mass Transfer in Biological Systems
108 Food and Bioengineering—Fundamental and Industrial Aspects
114 Advances in Bioengineering
163 Water Removal Processes: Drying and Concentration of Foods and Other Materials
172 Food, pharmaceutical and bioengineering—1976/77
181 Biochemical Engineering Renewable Sources of Energy and Chemical Foodstocks

CRYOGENICS

224 Cryogenic Processes and Equipment 1982
251 Cryogenic Properties, Processes and Applications 1986

CRYSTALLIZATION

110 Factors Influencing Size Distribution
193 Design Control and Analysis of Crystallization Processes
215 Nucleation, Growth and Impurity Effects in Crystallization Process Engineering
240 Advances in Crystallization From Solutions
253 Fundametnal Aspects of Crystallization and Precipitation Processes

DRAG REDUCTION

11 Drag Reduction
130 Drag Reduction in Polymer Solutions

ENERGY

Conversion and Transfer

5 Heat Transfer, Atlantic City
57 Heat Transfer, Boston
59 Heat Transfer, Cleveland
75 Energy Conversion Systems
79 Heat Transfer with Phase Change
87 Advances in Cryogenic Heat Transfer
113 Convective and Interfacial Heat Transfer
118 Heat Transfer—Tulsa
119 Commercial Power Generation
138 Heat Transfer—Research and Design
162 Energy and Resource Recover from Industrial and Municipal Solid Wastes
174 Heat Transfer, Research and Application
189 Heat Transfer—San Diego 1979
202 Transport with Chemical Reactions
208 Heat Transfer—Milwaukee 1981
216 Processing of Energy and Metallic Minerals
225 Heat Transfer—Seattle 1983
236 Heat Transfer—Niagra Falls 1984
245 Heat Transfer—Denver 1985
257 Heat Transfer—Pittsburgh 1987
263 Heat Transfer—Houston 1988
269 Heat Transfer—Philadelphia 1989

Nuclear Engineering

53 Part XIII
56 Part XIV
95 Part XX
104 Part XXI
106 Part XXII
119 Commercial Power Generation
168 Heat Transfer in Thermonuclear Power Systems
169 Developments in Uranium Enrichment
191 Nuclear Engineering Questions Power Reprocessing, Waste, Decontamination Fusion
221 Recent Developments in Uranium Enrichment

ENVIRONMENT

78 Water Reuse
97 Water—1969
115 Important Chemical Reactions in Air Pollution Control
122 Chemical Engineering Applications of Solid Waste Treatment
124 Water—1971
126 Air Pollution and its Control
133 Forest Products and the Environment
137 Recent Advances in Air Pollution Control
139 Advances In Processing and Utilization of Forest Products
144 Water—1974: I. Industrial Wastewater Treatment
145 Water—1974: II. Municipal Wastewater Treatment
146 Forest Product Residuals
147 Air: I. Pollution Control and Clean Energy
148 Air: II. Control of NO_{xx} and SO_x Emissions
149 Trace Contaminants in the Environment
151 Water—1975
156 Air Pollution Control and Clean Energy
157 New Horizons for the Chemical Engineer in Pulp and Paper Technology
165 Dispersion and Control of Atmospheric Emissions, New-Energy-Source Pollution Potential
170 Intermaterials Competition in the Management of Shrinking Resources
171 What the Filterman Needs to Know About Filtration
175 Control and Dispersion of Air Pollutants: Emphasis on NO_x and Particulate Emissions
177 Energy and Environmental Concerns in the Forest Products Industry
184 Advances in the Utilization and Processing of Forest Products
188 Control of Emissions from Stationary Combustion Sources Pollutant Detection and Behavior in the Atmosphere
195 The Role of Chemical Engineering in Utilizing the Nation's Forest Resources
196 Implications of the Clean Air Amendments of 1977 and of Energy Considerations for Air Pollution Control
198 Fundamentals and Applications of Solar Energy
200 New Process Alternatives in the Forest Products Industries
201 Emission Control from Stationary Power Sources: Technical, Economic and Environmental Assessments
207 The Use and Processing of Renewable Resources—Chemical Engineering Challenge of the Future
209 Water—1980
210 Fundamentals and Applications of Solar Energy II
211 Research Trends in Air Pollution Control: Scrubbing, Hot Gas Clean-up, Sampling and Analysis
213 Three Mile Island Cleanup
223 Advances in Production of Forest Products
232 Applications of Chemical Engineering in the Forest Products Industry
239 The Impact of Energy and Environmental Concerns on Chemical Engineering in the Forest Products Industry
243 Separation of Heavy Metals and Other Trace Contaminants
246 Advances in Process Analysis and Development in the Forest Products Industries.
265 Resource Recovery of Municipal Solid Wastes

FLUIDIZATION

101 Fundamental Processes in Fluidized Beds
105 Fluidization Fundamentals and Application
116 Fluidization: Fundamental Studies Solid-Fluid Reactions, and Applications
176 Fluidization Application to Coal Conversion Processes
205 Recent Advances in Fluidization and Fluid-Particle Systems
234 Fluidization and Fluid Particle Systems: Theories and Applications
241 Fluidization and Fluid Particle Systems: Recent Advances
255 New Developments in Fluidization and Fluid-Particle Systems
262 Fluidization Engineering: Fundamentals and Applications
270 Fluidization and Fluid Particle Systems: Fundamentals and Application

HISTORY OF CHEMICAL ENGINEERING

100 The History of Penicillin Production
235 Diamond Jubilee Historical/Review Volume

ION EXCHANGE

79 Adsorption and Ion Exchange Separations
219 Recent Advances in Adsorption and Ion Exchange
230 Adsorption and Ion Exchange—'83
233 Adsorption and Ion Exchange—Progress and Future Prospects
259 Recent Progress in Adsorption and Ion Exchange

KINETICS

25 Reaction Kinetics and Unit Operations
73 Kinetics and Catalysis

MINERALS

15 Mineral Engineering Techniques
85 Fossil Hydrocarbon and Mineral Processing
173 Fundamental Aspects of Hydrometallurgical Processes
180 Spinning Wire from Molten Metals
216 Processing of Energy and Metallic Minerals

PETROCHEMICALS

49 Polymer Processing
127 Declining Domestic Reserves—Effect on Petroleum and Petrochemical Industry
135 The Petroleum/Petrochemical Industry and the Ecological Challenge
142 Optimum Use of World Petroleum
212 Interfacial Phenomena in Enhanced Oil Recovery

PETROLEUM PROCESSING

103 C_4 Hydrocarbon Production and Distribution
127 Declining Domestic Reserves—Effect on Petroleum and Petrochemical Industry
135 The Petroleum/Petrochemical Industry and the Ecological Challenge
155 Oil Shale and Tar Sands
226 Underground Coal Gasification: The State of the Art

PHASE EQUILIBRIA

2 Pittsburgh and Houston
6 Collected Research Papers
88 Phase Equilibria and Gas Mixtures Properties

PROCESS DYNAMICS

36 Process Dynamics and Control
46 Process Systems Engineering
55 Process Control and Applied Mathematics
214 Selected Topics on Computer-Aided Process Design and Analysis
267 Process Sensing and Diagnostics

SEPARATION

120 Recent Advances in Separation Techniques
192 Recent Advances in Separation Techniques—II
250 Recent Advances in Separation Techniques—III

SONICS

109 Sonochemical Engineering

MISCELLANEOUS

48 Chemical Engineering Reviews
70 Small-Scale Equipment for Chemical Engineering Laboratories
112 Engineering, Chemistry, and Use of Plasma Reactors
125 Vacuum Technology at Low Temperatures
143 Standardization of Catalyst Test Methods
182 Biorheology
183 The Modern Undergraduate Laboratory Innovative Techniques
185 Electro Organic Synthesis Technology
186 Plasma Chemical Processing
187 Chronic Replacement of Kidney Function
194 Hazardous Chemical—Spills and Waterborne Transportation
203 A Review of AIChE's Design Institute for Physical Property Data (DIPPR) and Worldwide Affiliated Activities
204 Tutorial Lectures in Electrochemical Engineering and Technology
206 Controlled Release Systems
217 New Composite Materials and Technology
220 Uncertainty Analysis for Engineers
228 Problem Solving
229 Tutorial Lectures in Electrochemical Engineering and Technology—II
231 Data Base Implementation and Application
237 Awareness of Information Sources
238 New Developments in Liquid-Liquid Extractors: Selected Papers From ISEC '83
244 Experimental Results from the Design Institute for Physical Property Data. I: Phase Equilibria
247 Chemical Engineering Data Sources
248 Industrial Membrane Processes
249 Measurement of High Temperatures in Furnaces and Processes
252 Thin Liquid Film Phenomena
254 Electrochemical Engineering Applications
256 Experimental Results From the Design Institute for Physical Property Data: Phase Equilibria and Pure Component Properties
258 Fiber Optics: Processing and Applications
261 New Membrane Materials and Processes for Separation
266 Diffusion and Convection in Porous Catalysts
268 Membrane Reactor Technology
271 Experimental Results From the Design Institute for Physical Property Data: Phase Equilibria and Pure Component Properties—Part II.
272 Membrane Separations in Chemical Industries

MONOGRAPH SERIES

3 The Manufacture of Nitric Acid by the Oxidation of Ammonia—The DuPont Pressure Process by Thomas H. Chilton
4 Experiences and Experiments with Process Dynamics by Joel O. Hougen
5 Present, Past, and Future Property Estimation Techniques by Robert C. Reid
6 Catalysts and Reactors by James Wei
7 The 'Calculated' Loss-of-Coolant Accident by L.J. Ybarrondo, C.W. Solbrig, H.S. Isbin
8 Understanding and Conceiving Chemical Process by C. Judson King
9 Ecosystem Technology: Theory and Practice by Aaron J. Teller
10 Fundamentals of Fire and Explosion by Daniel R. Stull
11 Lumps, Models and Kinetics in Practice by Vern W. Weekman, Jr.
12 Lectures in Atmospheric Chemistry by John H. Seinfeld
13 Advanced Process Engineering by James R. Fair
14 Synfuels from Coal by Bernard S. Lee
15 Computer Modeling of Chemical Processes by J.D. Seader
16 "High-Tech" Materials by Sheldon Isakoff
17 Separations: New Directions for an Old Field
128 Biotechnology: Status and Perspectives

ISBN 0-8169-0482-0